MASTERING the ART of PROJECT MANAGEMENT ENGINEERING

All You Need to Know to be an Expert in Engineering Project Management

A Guide to Thinking beyond the Cloud as a Project Manager Engineer

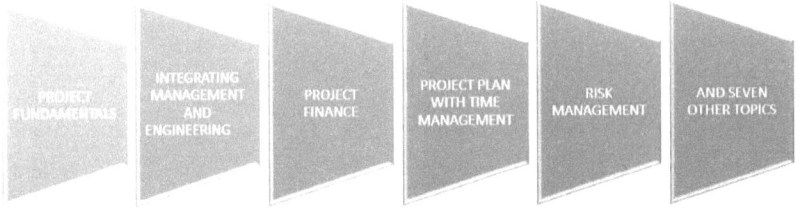

Oluwaseun Adenigba | *FAAPM®, MNSE, MSc, BEng, Dip.*

Copyright 2018 by Oluwaseun Adenigba. All rights reserved. No part of this publication may be reproduced, distributed, or transmitted in any form or by any means, including photocopying, recording, or other electronic or mechanical methods, without the prior written permission of the publisher, except in the case of brief quotations embodied in critical reviews and certain other noncommercial uses permitted by copyright law. For permission requests, write to this address: adenigbaoa@icloud.com

The author makes no representation, express or implied, regarding the accuracy of the information in this book and cannot accept any legal responsibility for any error or omissions that may be made.

Contents

Contents 3
INTRODUCTION 7
   Why this book is for you 7
   Who should read this book? 7
   Real life applications of PME 7
   What Project Management Engineering (PME) entails 8
   Case study outline 8
1.0 PROJECT FUNDAMENTALS 12
   1.1 Definition of a Project 12
   1.2 Project Characteristics 12
   1.3 Project Life Cycle 13
   1.4 Project Organizations 14
   1.5 Project Planning 15
   QUESTIONS 16
   ANSWERS 16
2.0 INTEGRATING MANAGEMENT AND ENGINEERING 18
   2.1 The Role of the Project Manager 18
   2.2 Purpose of Project Management 21
   2.3 Engineering Management 22
   2.4 Engineering Project Management 23
   QUESTIONS 24
   ANSWERS 24
3.0 DESIGN PROPOSALS 26
   3.1 Evolution of Projects 26
   3.2 Project Execution Plan 27
   3.3 Project Definition 27
   3.4 Design Proposals 30
   3.5 Engineering Project Controls 31
   QUESTIONS 32
   ANSWERS 32

4.0 DESIGN COORDINATION 34
4.1 Team Management 34
4.2 Design Effectiveness Evaluation 35
4.3 Constructability 41
4.4 Post Design Review 42
QUESTIONS 43
ANSWERS 43

5.0 PROJECT FINANCE 45
5.1 Funding for Budgets 45
5.2 Financial Sources 46
5.3 Project Finance 47
5.4 Financial Instruments 48
5.5 Financial Engineering 50
5.6 Debt Financing Contracts 51
5.7 Appraisal and Validity of financing projects 53
5.8 Risks 54
5.9 Financial Risks 55
QUESTIONS 59
ANSWERS 59

6.0 PROJECT PLAN WITH TIME MANAGEMENT 61
6.1 Project Plan 61
6.2 Developing a Project Network Plan 62
6.3 Activity-On-Node Network Techniques 66
6.5 Scheduling Techniques 79
6.6 Time Management 81
QUESTIONS 84
ANSWERS 84

7.0 STRATEGIC CONTRACT SELECTION 86
7.1 What is a Contract? 86
7.2 Right Contract Selection 88
7.3 Contract Types 90
7.4 Contract Documents 93
7.5 Construction claims 96

- 7.6 Selecting the Contractor 97
- 7.7 Sub-Contracting 98
  - QUESTIONS 100
  - ANSWERS 100
- **8.0 PROJECT MANAGEMENT AND QUALITY** 102
  - 8.1 Quality Concepts 102
  - 8.2 Project Quality Management 108
  - 8.3 Project Quality Management Processes 109
  - 8.4 Techniques for Quality Control 116
  - 8.5 Business process re-engineering (BPR) and Total Quality Management (TQM) 122
  - QUESTIONS 124
  - ANSWERS 124
- **9.0 RISK MANAGEMENT** 126
  - 9.1 Introduction 126
  - 9.2 Risk Management 127
  - 9.3 Risk Identification 130
  - 9.4 Risk Analysis 133
  - 9.5 Risk Response 135
  - 9.6 Risk Control 137
  - QUESTIONS 139
  - ANSWERS 139
- **10.0 STAKEHOLDER MANAGEMENT** 141
  - 10.1 Introduction 141
  - 10. 2 Primary Project stakeholders 141
  - 10.3 Secondary project stakeholders 143
  - 10.4 Understanding the interests and influences 144
  - 10.5 Managing stakeholders 145
  - 10.6 Stakeholders and communication 148
  - QUESTIONS 150
  - ANSWERS 150
- **11.0 PROJECT CLOSURE** 152
  - 11.1 Testing and Start-Up 152
  - 11.2 Final Inspection 153

11.3 Guarantee and Inspections 155

11.4 Planned Drawings and Records 155

11.5 Disposition of Project Plans 156

11.6 Post Project Critique 156

11.7 Client feedback 157

    QUESTIONS 158

    ANSWERS 158

12.0 FUTURISTIC VIEW 160

    12.1 ENGINEERING PROJECT MANAGEMENT – THE WAY FORWARD 160

REFERENCES 162

About the Author | Oluwaseun Adenigba 163

Detail Qualifications 166

Past Projects by the Author 168

INTRODUCTION

Why this book is for you

This book will help maximize your budget to achieve quality projects

Improve effectiveness and increase the development of the project team.

It would also improve risk assessment in carrying out projects.

Who should read this book?

The following people will find this book beneficial and appropriate

- Students
- Engineers involved in projects
- Project Managers at professional level
- Project team members
- Project stakeholders and project customers
- Consultants and experts in project management
- Trainers developing project management educational courses
- Senior executives and Directors

Real life applications of PME

These industries engage PME in real life

- Engineering and Construction
- Oil and Gas Sector
- Information Technology
- Energy Sector
- Manufacturing Industry
- Agriculture
- Estate Management
- FMCGs
- Healthcare
- Many more

What Project Management Engineering (PME) entails

PME involves theoretical and practical characteristics of small and large engineering project management. This book provides the techniques and principles of project management (PM) starting with project management fundamentals.

Other chapters include effective project coordination, contract in PM, time management, risk and finance management and closing in on futuristic perception of PME.

The key to having a successful project relies on selecting and coordinating people that have the skills to detect, analyze and solve problems to complete the project. Hence, the book consists of relevant information to achieve this.

The chapters also involve the engineer's perspective thus giving the book an additional advantage when compared to the conventional project management books

Various tables and charts are displayed and highlighted in this book to enhance the understanding of the reader particularly the three essential segments of a project: scope, budget and schedule.

All through this book, accomplishing project quality to meet the owner's objective is emphasized as an essential piece of project management.

Case study outline

Bill and Mark (recently joined managers) have been assigned to Paul (Senior Manager) in the Management Consulting Office for six months coaching period on Engineering Project Management techniques relevant to their clients.

Bill and Mark are so excited to have their project management engineering skills sharpened by Paul as they discovered a few days earlier that He is one of the best senior managers in the company who is well respected.

Hello, Bill and Mark greet Paul as they enter the conference room.

Please take your seat, Paul replied.

Can anyone volunteer to tell us why we are here, questioned Paul

Bill responds with excitement but in a low voice, "we are here to learn about the "Engineering Project Management."

That's correct replied Paul.

Any other reason questioned Paul as he beckons on Mark

Mark responds "we are here to be masters in engineering project management."

That is also true says Paul, and went to the white board near the TV screen in the room.

I think it's better if I give an overview of the session. I will highlight the main sections of engineering project management covering significant areas that would instantly boost off your knowledge, and I will go into details for each area as we progress. I am sure you will enjoy this session.

Bill wrote the following points:

PROJECT FUNDAMENTALS	INTEGRATING MANAGEMENT AND ENGINEERING	DESIGN PROPOSALS	DESIGN COORDINATION
PROJECT FINANCE	PROJECT PLAN WITH TIME MANAGEMENT	STRATEGIC CONTRACT SELECTION	PROJECT MANAGEMENT AND QUALITY
RISK MANAGEMENT	STAKEHOLDER MANAGEMENT	PROJECT CLOSURE	FUTURISTIC VIEW

CHAPTER 1

PROJECT FUNDAMENTALS

- Definition of a Project
- Project Characteristics
- Project Life Cycle
- Project Organization
- Project Planning

1.0 PROJECT FUNDAMENTALS

1.1 Definition of a Project

Ok, let's begin by defining what a Project is.

Mark positioned his chair nearer to the table to take notes.

Paul wrote,

A project consists of interconnected work activities constrained by a specific schedule, budget, and scope to deliver capital assets needed to achieve the strategic purpose of an organization or meet the needs of our clients.

This could probably mean that projects are temporal whispered Mark.

That's true replied Paul, as he turned towards them smiling and Bill nodded in agreement

1.2 Project Characteristics

Paul continued, now I will list out the characteristics of a Project

(John M. Nicholas and Herman Steyn, 2012)

- Projects must be well defined
- Projects have a definite goal
- Projects require resources to be accomplished
- Projects are time bound i.e. have a scheduled time frame
- Projects are accomplished in phases
- Projects are undertaken to generate expected results

What if we have an idea but haven't got the best way to define our goal specifically, could that still be termed a project, questioned Bill

No, at all, answered Paul.

Paul further said, "Your idea must be translated into a definite objective or goal with the necessary resources to carry it out before we can safely conclude that a framework of a project is being formed."

Hmmm, I see, said Bill

1.3 Project Life Cycle

So, at this moment, said Paul, I will show you the major life cycle of the Engineering Project Management.

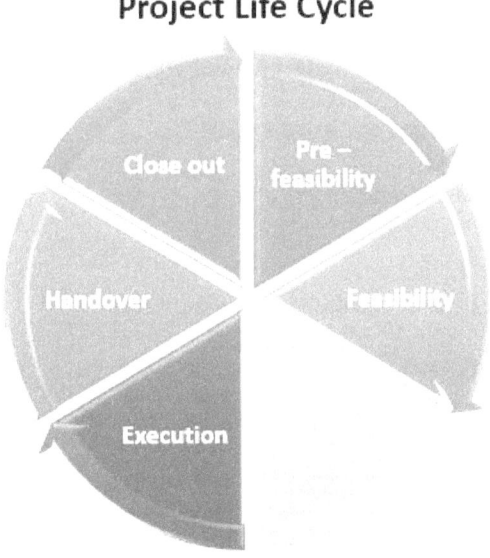

1. Pre – feasibility: This is where the project need is identified, and initial validation of concepts options are done
2. Feasibility: A thorough feasibility investigation is done here including preliminary brief and analysis of project investment and estimate.
3. Planning: At this third stage, the project is defined concerning scope, budget, control procedures and schedule.
4. Execution: The project is implemented in phases at this stage
5. Handover: This stage comprises of passing the facilities, user training, operating and maintenance documentation and possible technical know-how to the client or end user depending on the project type
6. Close out: The close out stage brings the project to an end by transferring lessons learned, establishing appropriate performance evaluations, keeping projects records and dispatching project team

Quite interesting, Bill commented

I suggest we term it as the 'PFPIEC' life cycle, pronounced 'pif-pik' Mark suggested,

Or "The 6 LC's" where LC signifies Life Cycle Bill said,

Why not, said Paul if it enhances your understanding.

1.4 Project Organizations

So, we shall be looking at project organization next, said Paul

The implementation and development of a project influence the structure of an organization. Many organizations function by a staff reporting to his superior.

For your information, Paul spoke motioning to Bill and Mark to have their attention,

Our Management Consulting Company does not function that way, we operate in teams, and that is why this coaching class consist of a team of you both.

In the same vein, a project-oriented organization establishes teams to implement a project. This team would have a project manager not to be a "lord" over them but to lead, guide and direct team ideas, and resources more appropriately

I think I would fit in this organization since I love teamwork, Bill commented.

Paul smiled.

1.5 Project Planning

Lastly, for today's coaching, we shall be looking at how Engineering Project Management is planned, said Paul as he took a sip of water from his bottle.

Effective planning is key to succeeding in Engineering Project Management. It involves a strategy to carry out tasks promptly.

It includes

1. Specifying the roles and responsibilities of the project team
2. Scheduling for design
3. Scheduling for production, construction or manufacturing as the case may be
4. Procurement planning as well

That's the end for today's coaching, see you tomorrow stated Paul.

QUESTIONS
CHAPTER 1: INTRODUCTION

1. A project requires little or no resources to be accomplished
   a) True   b) False
2. Executing engineering projects are done in phases
   a) True   b) False
3. The life cycle of a project ends with a .... Stage
   a) Handover b) feasibility c) close out
4. Project development and implementation influences the structure of an organization a little
   a) True   b) False
5. Scheduling for .... and .... Is part of the project planning process
   a) Manufacturing and Gist   b) design and manufacturing   c) scheduling and meetings

ANSWERS
1. B   2. A   3. C   4. B   5. B

CHAPTER 2

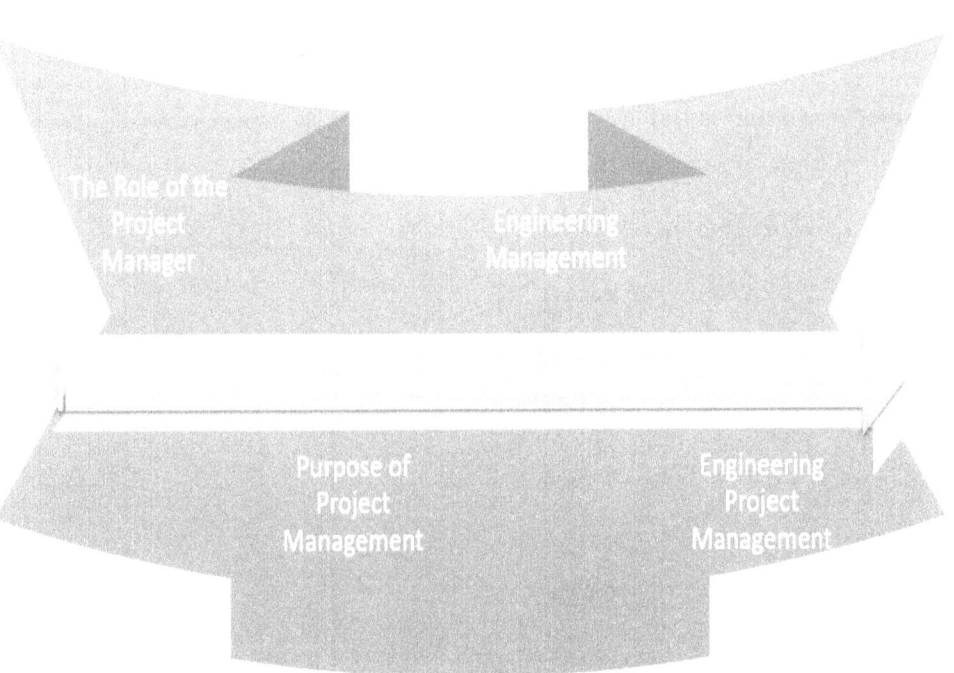

- The Role of the Project Manager
- Engineering Management
- Purpose of Project Management
- Engineering Project Management

2.0 INTEGRATING MANAGEMENT AND ENGINEERING

The next day

Hi, Bill and Mark hails Paul as they both take their seat positioning themselves to face the board.

Paul smiled in affirmation and said

Our next line of discussion will reveal the relevance of combining management and engineering skills to be a master in Engineering Project Management, Paul stated

We shall be looking at the following five topics

1. The Role of the Project Manager
2. Purpose of Project Management
3. Engineering Management
4. Engineering Organization
5. Engineering Project Management

2.1 The Role of the Project Manager

Firstly, let us discuss the role of the Project Manager

A project manager is to execute quality projects by leading the project team within the planned budget, time and scope.

Your responsibilities as a Project Manager fall into four major categories (Heerkens, 2002)

He must adequately manage problems and risks that may arise during projects so he can deliver the results

He must carry out required duties in logical sequence and maximize available resources to the best advantage

The Project Manager must focus on the five basic management functions which are

Planning

Controlling   Organizing

Directing   Staffing

These five basic functions must be sequential in their approach to managing engineering projects, questioned Bill

Yes Bill, replied Paul

Any ideas why this is necessary, as Paul looks at both Mark and Bill

Hmm, I suppose it is to ensure that the project is adequately managed to minimize waste of resources and to deliver quality results I guess, answered Mark

You are right, stated Paul

So, I will highlight in a table format for easy understanding the project manager role in these five functions

Role of the Project Manager in:

Planning
1. Establishes the aim of the project early, so the project team is aware of the requirements
2. Factors in possible contingencies in the planning phase and schedule a reserve for possible future problems
3. Involves key staff members and assistant managers in the planning process
4. Prepares official agreements with appropriate parties during a project change and establish systems to control such change
5. Communicates project plan specifying individual responsibilities, budgets and schedules

Organizing
1. Organizes the project around the work to be implemented
2. Develops a work breakdown structure that divides the project into specific and measurable units
3. Establishes a project organization chart for each project spelling out what is to be done and by who
4. Defines the duties and authority of all project team members

Staffing
1. Orientates team members on project goals and objectives at the inception of project
2. Defines task to be carried out, and work with appropriate department managers in selecting project team members
3. Collects individual member's input to agree upon scope budget and schedule
4. Explains the project expected to team members and the impact of their contribution on the project

Directing
1. Displays effective leadership in coordinating all important aspects of the project
2. Displays high-level interest and zeal in the project with an optimistic attitude
3. Is easily accessible by team members to guide, assist them in cooperatively implementing project task
4. Investigates and analyzes problems early enough to provide solutions more quickly

5. Recognizes the relevance of team members, publicly acknowledge the good work, guide them in correcting mistakes, and build an effective team

Controlling
1. Measures project plan and actual accomplishment and maintain correlation.
2. Maintains an updated milestone chart that displays planned and achieved milestones
3. Archives records of meetings, telephone conversations, and agreements
4. Continuously updates his team on necessary information and proposes solutions to problems

Wow, so many responsibilities of a project manager commented Bill

True, that is why our Management Consulting Company does not compromise on the expectations of our Engineering Project Managers. Thus, we have produced outstanding managers over the years

I agree, stated Mark.

2.2 Purpose of Project Management

It was Myles Munroe who said "When purpose is not known, abuse is inevitable" therefore, we shall be discussing the purpose of Project Management

Recently, managers have discovered that they are continuously involved in projects which exhibit cost overruns, time extensions, conflicts, and negotiation management. It is for this purpose that engineering managers must be empowered with adequate project management skills to tackle the rising challenges in projects today.

Many project activities often require external individuals outside of the project manager's organization. Although these individuals are not answerable to the project manager directly, the knowledge of project management helps the manager develop effective working relationships.

Project Management knowledge gives the management an edge to accurately answer

Who does what?

When?

How much?

Lastly, on our discussion on the Purpose of Project Management, Project Management helps the manager to adequately coordinate people, equipment, material and financial resources to achieving a specified project on time and within approved cost

I am enjoying this coaching session, learning a lot stated Mark

I'm glad you do, replied Paul

2.3 Engineering Management

So, in this subtopic, we will be conversing on Engineering Management

Engineering Management is a specialized management system that deals with the application of engineering principles and technicality. It is concerned with management in the engineering field including but not limited to production, manufacturing, construction, design, and areas where engineering functions are carried out.

Engineering Project managers combine their skills and experience in business and engineering. They can maximize machines and human resources to achieve project objective.

To be successful Project Management Engineers, you will learn to master people skills and be competent in engineering techniques.

This is quite a lot to learn Paul, stated Bill

You are right, although it gets easier as you gain experience in managing projects, replied Paul

I think a "can do" attitude will help to achieve this, suggested Mark.

Absolutely, stated Paul

2.4 Engineering Project Management

We shall be discussing our last topic for chapter 2 which is Engineering Project Management, stated Paul

The demand for a better lifestyle and advancement in the engineering sector combined with the arrival of technological innovation and the internet has led to an increase in a complex number of large engineering projects. The unique feature of each project makes successful project management a cumbersome task. These projects are usually involved with significant risk and engineering implication.

Engineering Project Management specifically applies engineering management to project activities.

It is the specific application of engineering management to completing a specific project on time and within approved cost.

Wow, this is worth noting down, Mark remarked

Indeed, that's the objective of this coaching class, replied Paul

QUESTIONS
CHAPTER 2: INTEGRATING MANAGEMENT AND ENGINEERING

1. The project, organization team and .... are the major responsibilities of the project manager
   a) Client b) contractor c) yourself d) government

2. .... is listed among the basic functions of the project manager
   a) Meditating b) Meetings c)Lunching d)Staffing

3. Competence in engineering techniques and mastering people skills aids not the success of the project manager engineer
   a) True b) false c) no idea

4. Engineering Project Management specifically applies engineering management to project activities.
   a) True b) false c) no idea

5. Maintaining an updated milestone chart that displays planned and achieved milestones is part of which of the project manager functions?
   a) Directing b) Controlling c) Staffing d) Planning

ANSWERS
1. C   2. D   3. B   4. A   5. B

CHAPTER 3

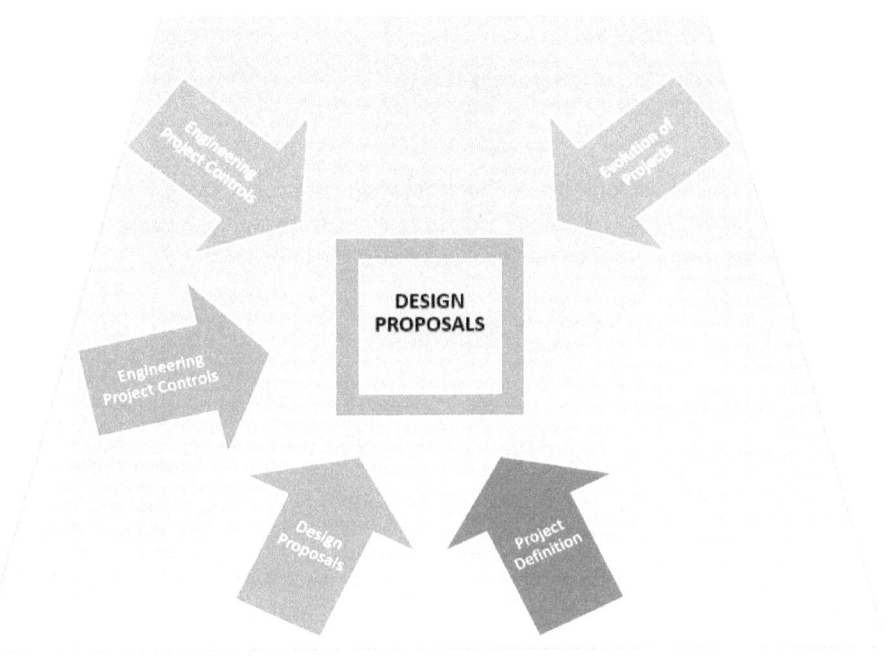

3.0 DESIGN PROPOSALS

So, for this chapter on Design Proposals, I will cover the following areas

1. Evolution of Projects
2. Work Plan Design Development
3. Project Controls in Engineering
4. Progress Measurement of Engineering Design

3.1 Evolution of Projects

It is of high relevance that the Engineering Project Manager is involved in the design effort of the project. He must be aware that as the project progresses from initiation to closeout phase, it experiences gradual change through the following stages (Oberlender, 2000).

Client's development stage
Organization stage
Engineering stage
Procurement stage
Construction stage
System testing and start-up phase
Project completion and contract close-out phase

Request for Proposal (RFP) to further develop the project is made available at the end of Client's development stage.

It is of interest to note that our Management Consulting Company is passionate about delivering excellent results to our client. Therefore, we ensure a clear understanding of our client's proposal request and goal.

It is expedient for us especially the assigned project manager engineer to effectively grasp the desired outcome of the project and the reasons the project is being undertaken by the client.

Never forget this Bill, as Paul turned to him

Always remember this Mark, as Paul motioned to him

Both nodded in agreement

3.2 Project Execution Plan

The development of a project execution plan to manage the design process is the first step to developing a design proposal. The plan consists of the work scope covered in the RFP and interfaces with other team members or outsiders involved in the project.

The Project manager must ensure that the plan must comprise of milestone schedule which shows major stages, areas of work and critical due dates. Also, an overall preliminary budget must be developed to serve as a guide for the project to cater for unforeseen surprises as the project develops.

History reveals that many project managers have expressed frustration as the project is implemented simply due to vaguely defined scopes of work with RFPs.

3.3 Project Definition

We understand that the answers to questions such as

1. What are we trying to do?
2. What do we know about the project?
3. What work do we need to do?

...are prerequisites to having a clear project definition.

It is difficult to outline the scope of design work without a properly defined project.

Project definition is a criterion to effective engineering design.

A poorly defined project scope results in major project changes, rework cost overruns and schedule delays.

The Project Definition Rating Index (PDRI) below serves as a guide to measure the level of definition of a project. It allows the project engineer and his team to rate, enumerate and access the scope development level for projects before approval for detailed design and construction.

Project Definition Rating Index (PDRI)--Sections, Categories, and Elements.

Source: *Construction Industry Institute.*

	1		2		3
	Basis of Project Decision		**Front End Definition**		**Execution Approach**
A	Manufacturing Objectives Criteria	F	Site Information	L	Procurement Strategy
A1	Reliability Philosophy	F1	Site Location	L1	Procurement Strategy Equipment & Materials
A2	Maintenance Philosophy	F2	Surveys & Soil Tests	L2	Procurement Procedures & Plans
A3	Operating Philosophy	F3	Environmental Assessment	L3	Procurement Responsibility Matrix
		F4	Permit Requirements	M	Deliverables
B	Business Objectives	F5	Utility Sources with Supply Conditions	M1	CADD/Model Requirements
B1	Products	F6	Fire Protection & Safety Considerations	M2	Deliverable Defined
B2	Market Strategy	G	Process / Mechanical	M3	Distribution Matrix
B3	Project Strategy	G1	Process Flow Sheets	N	Project Control
B4	Affordability or Feasibility	G2	Heat & Material Balances	N1	Project Control Requirements
B5	Capacities	G3	Piping & Instrument Diagrams (P&IDs)	N2	Project Accounting Requirements
B6	Future Expansion Considerations	G4	Process Safety Mgmt (PSM)	N3	Risk Analysis
B7	Expected Project Life Cycle	G5	Utility Flow Diagrams	P	Project Execution Plan
B8	Social Issues	G6	Specifications	P1	Owner / Approval Requirements
C	Basic Data Research & Development	G7	Piping System Requirements	P2	Engineering / Construction Plan & Approach
C1	Technology	G8	Plot Plan	P3	Shut Down / Turn-Around Requirements
C2	Processes	G9	Mechanical Equipment List	P4	Pre-Commissioning Turnover Sequence Requirements

D	Project Scope	G10	Line List	P5	Startup Requirements
D1	Project Objectives Statement	G11	Tie-in List	P6	Training Requirements
D2	Project Design Criteria	G12	Piping Specialty Items List		
D3	2	H	Equipment Scope		
D4	Dismantling & Demolition Requirements	H1	Equipment Status		
D5	Lead / Discipline Scope of Work	H2	Equipment Location Drawing		
D6	Project Schedule	H3	Equipment Utility Requirements		
E	Value Engineering	I	Civil, Structural, & Architectural		
E1	Process Simplification	I1	Civil / Structural Requirements		
E2	Design & Material Alternatives Considered / Rejected	I2	Architectural Requirements		
E3	Design for Constructability Analysis	J	Infrastructure		
		J1	Water Treatment Requirements		
		J2	Loading / Unloading / Storage Facilities Requirements		
		J3	Transportation Requirements		
		K	Instrument & Electrical		
		K1	Control Philosophy		
		K2	Logic Diagrams		
		K3	Electrical Area Classifications		
		K4	Substation Requirements (Power Sources Identified)		
		K5	Electric Single Line Diagrams		
		K6	Instrument & Electrical Specifications		

The PDRI (Project Definition Rating Index) is divided into 3 major sections with 15 subsections which consist of 70 elements.

To determine the PDRI, each element is rated on a scale from 1 – 5, where an element rating 1 signifies complete definition whereas a rating of 5 signifies incomplete or poor definition of an element.

The sum of the element weights is the composite weighted score of a project, which could range up to 1,000 points, with lower points signifying better score and higher points showing a worse score.

This makes a lot of sense, remarked Bill

Yes, the PDRI table easily shows us if a project is well defined or not, Bill commented

You both are right, Paul affirmed

3.4 Design Proposals

Moving forward, we shall discuss the Design Proposal and its impact in Engineering Project Management.

Project Manager should carefully review the request for proposal (RFP) upon receipt. This is to enable familiarity to information such as environmental and community relations, hazardous waste, bidding strategy, required permits and procedures, expectations and customer objectives.

In a situation where client or sponsor is unable to clarify or define proposal fully, the engineering project manager who may likely be the design engineer must define the scope of engineering task to the best of his or her ability and develop a budget and schedule based on the assumed scope of work.

When this is achieved, it is expedient to document and communicate to the client the assumptions made and the possible impact of the work on the total project.

Any ideas why this documentation and communication is necessary, questioned Paul as he turned to face both Bill and Mark

This keeps client "in the know" replied Bill

And to ensure that client's objective is realized says, Mark

You both are right, stated Paul

Also, this essentially locks in the scope of work at this phase in the project. Additional information such as assumed scope, budget, and schedule for that portion could be added as the project progresses.

3.5 Engineering Project Controls

A process to control scope change(s) must be in place in any design effort. This ensures that client and project team members are carried along on the impact of change on project cost and schedule including effective communication between project manager and client.

Also, a system to measure progress and control schedule must be established. This should consist of Work Breakdown Structure (WBS) for engineering, engineering manager, and teams' roles as well as an external consultant or personnel roles. A system for cost management and control must be added at the design stage for measuring productivity, report cost performance and can be utilized to obtain change approval in the engineering budget.

QUESTIONS
CHAPTER 3: DESIGN PROPOSALS

1. The initial step to developing a project execution plan is ....
   a) Design strategy b) Design proposal c) Design execution

2. It is easy to outline the scope of design work without an adequate defined project
   a) True b) False c) No idea

3. ....serves as a guide to measure the project definition level
   a) PDIR b) PDRI c) DPRI

4. Cost management and control system aids to measure productivity
   a) True b) False c) No idea

5. Equipment scope, infrastructure and site information are part of ....
   a) Front-end definition b) execution approach c) project decision

ANSWERS
1. B   2. B   3. B   4. A   5. A

CHAPTER 4

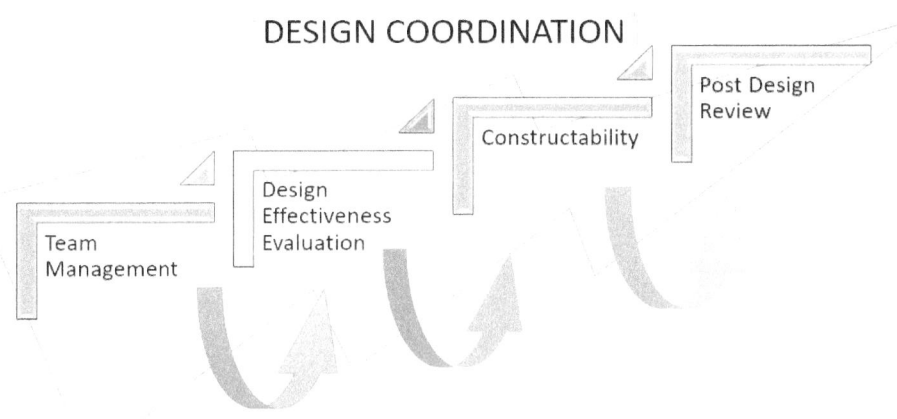

4.0 DESIGN COORDINATION

In this chapter on Design Coordination in Engineering Project Management, we shall focus on the following

1. Team Management
2. Design Effectiveness Evaluation
3. Constructability
4. Post Design Review

4.1 Team Management

A disorganized team cannot implement quality project plan neither can they maximize available resources effectively, therefore it is beneficial to have a coordinated team to achieve client's objective.

A team is a group of individuals, responsible for specific task(s)

It is the duty of the project manager to successfully coordinate the team to minimize conflicts, misunderstandings, delay and disruption among team members or between the client and the team.

The project manager engineer must quickly identify potential problems and respond rapidly to resolve challenges that may arise during project implementation

Common Problems in Team Management

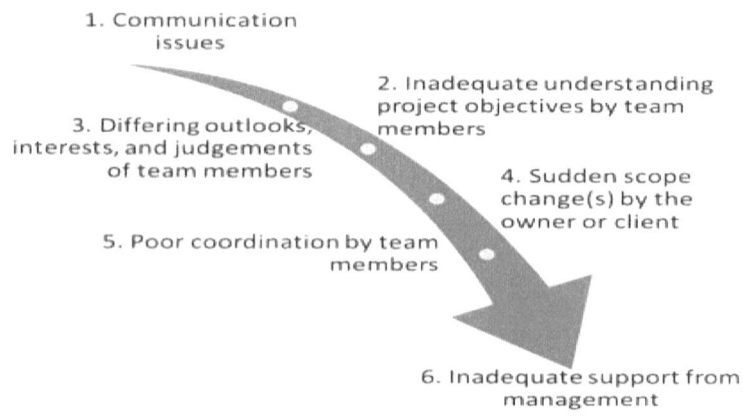

1. Communication issues
2. Inadequate understanding project objectives by team members
3. Differing outlooks, interests, and judgements of team members
4. Sudden scope change(s) by the owner or client
5. Poor coordination by team members
6. Inadequate support from management

Misplaced priorities can occur as the project unfolds and team members are assigned to handle more responsibilities. It is therefore of necessity that priorities be set at the beginning of the project. Objectives of a project must be understood by team members to avoid miscommunication and misunderstandings.

Furthermore, team meetings must regularly be held to share relevant and current information about recent events and to serve as a medium for tackling problems.

Project manager must also be accessible to team members who may feel uncomfortable airing suggestions publicly or potential problems in meetings.

The project manager is responsible for overall coordination of the project team. However, individual coordination is also necessary among team members. The project manager must inculcate an environment that fosters cooperation and encourages a free exchange of information among team members.

Each project participant should be made to understand the power of team accomplishment. The project manager minimizes conflicts by emphasizing commitment, clarity, and unity of the team. Also, the project manager must handle variation in priorities, interests, and judgments from the team skillfully.

Does this mean conflict must be present in all situations in managing engineering projects, questioned Mark?

Very likely, but the master project management engineer's goal is to minimize this to the barest minimum, replied Paul.

I see, commented Bill.

4.2 Design Effectiveness Evaluation

Design involves multidimensional processes to produce specific instructions for the implementation of the project. This is normally achieved by applying technical knowledge to innovative ideas.

A design is said to be highly effective in engineering project management if it maximizes the cost and schedule of the project. That is, it has a significant impact on project cost and schedule. All these complex processes are to ensure the most effective design possible.

Measuring design productivity could be more difficult when compared to measuring productivity at the construction stage. Simple measuring tools for example cost per drawing or work-hours per drawing considering varieties in content and drawing size.

Likewise, there is growing recognition that the genuine measures of the effectiveness of the design effort are found in the construction of a project by the client after construction is finished.

Therefore, it might be more valuable to develop a technique for assessing the effectiveness of design instead of the productivity during the design stage.

The capacity to quantify design effectiveness utilizing the proposed method signifies a vital step to improve the total design process.

Construction Industry Institute (CII) researchers have proposed a method termed 'Objectives Matrix for productivity Evaluation". This technique can also be used to develop an effective measurement for design.

This objective matrix has four major components

1. Criteria: Defines what is to be measured
2. Weights: Determines the relative importance of the criteria to each other and the overall objective of the measurement
3. Performance scale: It relates the measured value of the criterion to a standard or selected benchmark value
4. Performance Index: Indicates progress and tracks performance using the other three components

The diagrams below are excerpt summary from Construction Industry Institute (CII) which serve as a measure of design effectiveness and an evaluation of design outputs.

N.B The report states this method is not intended to be, nor should it be, used as an evaluation of a designer or the design process

Use of the objective matrix is illustrated below. The seven criteria of design effectiveness are given as column headings; the criteria weights are shown near the bottom of each column. The performance scale of 0 to 10 is given to the right of the matrix. A score of 10 represents perfection, and a score of 3 is average. A score for the seven criteria is represented by an 'X' in the appropriate box and is recorded at the bottom of the matrix.

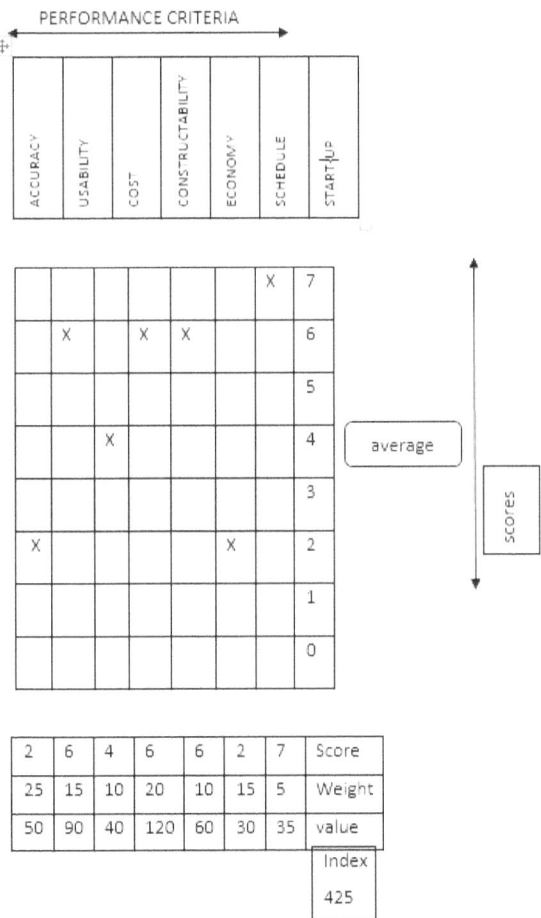

- Design Evaluation Matrix
- Source: Construction Industry Institute, Publication No. 8-1

Each criterion score is multiplied by its weight to obtain its value. The performance index displayed at the bottom right of the matrix is the sum of all values.

The score for a criterion can be obtained in at least three ways: judgmental, based upon a single quantitative measurement, or based upon a combination of several sub-criteria that are represented by a matrix. Judgmental scoring can be used for some or all criteria.

The scoring in the above figure shows judgmental scoring approach. Although this approach could be subjective, the objective matrix allows their appropriate application with multiple criteria of different weights.

For some criteria, quantitative measures can be used instead of judgments to determine scores. For instance, the accuracy of the design documents can be evaluated by measuring the amount of drawing revisions per total drawings.

The score for performance against the schedule criterion can be determined by using the percent of design document release dates attained. This approach is illustrated in Figure below, in which these two criteria are represented by quantitative measures.

In the example below, predetermined benchmark values are inputted into the boxes representing appropriate scores for each criterion.

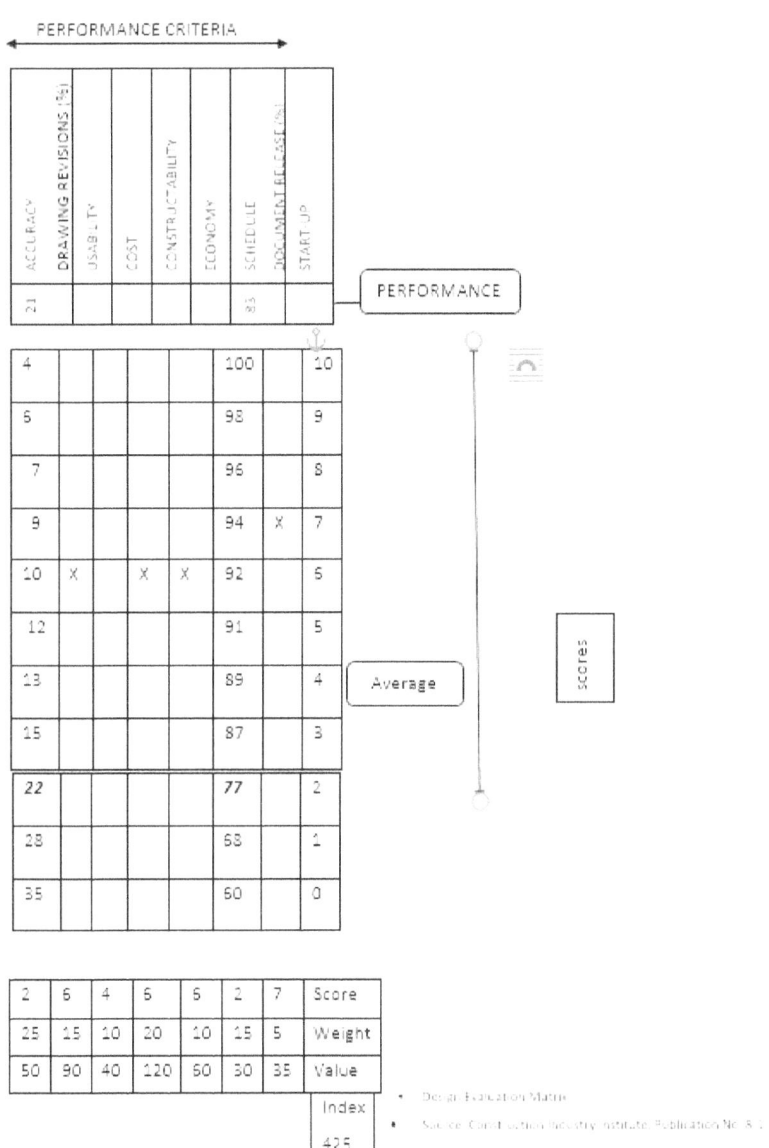

- Design Evaluation Matrix
- Source: Construction Industry Institute, Publication No. 8-1

The benchmark value for a score of 3 is considered average rather than 5 to give the opportunity for improvement, whereas a '0' represents the minimum level of accomplishment realized in recent years. The benchmark value for a level of 10 signifies the ultimate expectation in the foreseeable future.

The performance value attained is entered at the top of each column. The appropriate score representing that performance is then determined and is circled or "marked" in Figure above. A score is not given until it is attained and if a performance level falls between two scores, the lower score is used.

A third method is to measure each criterion using several sub-criteria. The sub criteria themselves can have differing weights and measurements. These can be combined into a single criterion score by a sub-matrix for that criterion.

Given the many intricacies and variables of the total design process, no measurement system can yield absolute quantitative results that are applicable without an interpretation of all the design circumstances.

However, the technique outlined can be utilized for three purposes: to develop a common understanding between the owner, designer, and contractor concerning the criteria by which design effectiveness on a given project will be measured.

Others involve comparing design effectiveness of similar projects in a systematic and reasonable quantitative manner, stressing performance trends; and to recognize opportunities to improve the effectiveness of the entire design process as well as the contributions toward the ultimate result made by all members.

Quite an eye opener strategy, commented, Mark.

Valid, says Paul.

4.3 Constructability

Paul continued,

The use of modern technology such as three-dimensional Computer Aided Drawing (CAD), automation and robotics in engineering, has produced increased interest in the constructability of a project. Designs can be easily configured more efficiently with these innovations. The outcome of this process and result facilitates exchange of ideas between construction, engineering, and design before and during design, rather than after design

The CII reports related to constructability. CII Publication No. 3-3, entitled Constructability Concepts File, provides a good description of constructability concepts related to conceptual planning, design and procurement, and field operations.

The following paragraphs contain excerpts from the report to illustrate the contents of the report.

There are at least five factors that should be considered in constructability deliberations related to design configurations for efficient construction:

1. Simplicity: Engineering and constructible designs must be simple enough, eliminating unnecessary complexities. Although special drawings and instruction may be required to improve the constructability process, particularly for retrofit or rebuild projects

2. Flexibility: This includes the use of innovative and alternative construction approaches by the engineering project manager to provide profitable design results

3. Sequencing: This is the procurement and construction of design consideration which allows for simultaneous multiple construction operations leading to an effective sequencing of project installation.

4. Substitution: It is the art of using alternatives constructability programs during the design phase to properly consider material applications that will positively impact constructability and reduce costly modifications

5. Labor skill/availability. The Project manager engineer must consider the availability of labor and the skill level of the workers early enough in the design phase of project life cycle

CII research reveals that the company or project size is no barrier to constructability and the implementation of a constructability program.

The involvement of construction in the design phase produces lower costs, improves productivity, results in earlier project completion and ultimately a better project. A key problem with the implementation of effective constructability programs is the concept that designs are selected to favour easiest to build ones.

CII Publication No. 3-2 provides guidelines for implementing a constructability program

4.4 Post Design Review

Our Consultancy firm, says Paul educate project managers to adopt a flexible approach to respond rapidly to change in individual projects. That is why our project managers are encouraged to ensure the determination of necessary modifications and improvements at the commencement of the project.

After completion of the design for each project, the project manager and his or her team should conduct a complete and candid evaluation of the design effort and the management of the design process.

This evaluation should include each member of the project team as well as other key participants that were involved in the design.

A checklist should be prepared to evaluate all aspects of the project. They include scope growth, match of quality and scope, owner's expectations and satisfaction.

Others are conflicts within the team or other parties, excessive changes in schedules, comparison of final costs with the original budget, and a list of precautions for management of future projects.

After thorough discussion of the design process, a summary report should be prepared by the project manager that should include a list of recommendations to improve the system for future projects.

Quite educative, remarked Mark.

QUESTIONS
CHAPTER 4: DESIGN COORDINATION

1. Sudden change of scope can cause problems in team management
   a) True b) false c) no idea

2. Individual coordination is not relevant as long as project manager effectively coordinates project team
   a) True b) false c) no idea

3. A design is highly effective in engineering project management if it maximizes the cost and schedule of the project
   a) True b) false c) no idea

4. The objective matrix has 4 major components Weights, performance scale, performance index and ....

5. ....and ....should be considered in constructability deliberations
   a) Sequencing and substitution b) elimination and substitution c) rigidity and flexibility

ANSWERS
1. A   2. B   3. A   4. Criteria   5. Sequencing and substitution

CHAPTER 5

PROJECT FINANCE

Funding for Budgets

Financial Sources

Project Finance

Financial Instruments

Financial Engineering

Debt Financing Contracts

Appraisal and Validity of financing projects

5.0 PROJECT FINANCE

So, in this Chapter, we shall take a look at the subtopics below:

Funding for Budgets

1. Project finance
2. Financial Sources
3. Financial Engineering
4. Financial Instruments
5. Debt Financing Contracts
6. Appraisal and Validity of financing projects
7. Risks

5.1 Funding for Budgets

Major construction projects in past five decades have been funded by public finance. One key channel to accomplish this is via taxation. Some or all the monies raised for projects such as transport systems, bridges and tunnels, motorways, power, and water plants flow from taxation.

In several cases, it has been a combination of a financial package of private sector loans and low-interest-rate loans or subordinated loans by the government. The need for private finance is due to the interest rate costs of borrowing money. The government must pay for these costs, therefore, warranting the aid of private finance to fund major projects.

Since the cost of finance and its related components are often determined by the type of project, its location and revenue generation method, the use of private finance rather than that of the public is justified if it offers a more profitable solution.

The impact of the financial plan of a project is often displayed in a greater way on the project success than the physical design or construction costs. The usual support by the government to project finance lenders in the usage of guarantees has endured the progress of projects in such way that would not have been commercially feasible without such provision.

5.2 Financial Sources

Finance for major projects is often made available by a lender such as a pension fund, an insurance company, a commercial bank, an export credit agency or a development bank. Additional sources of finance are large corporations, investment banks, institutional investors, niche banks and developers, vendors, contractors and utility subsidiaries.

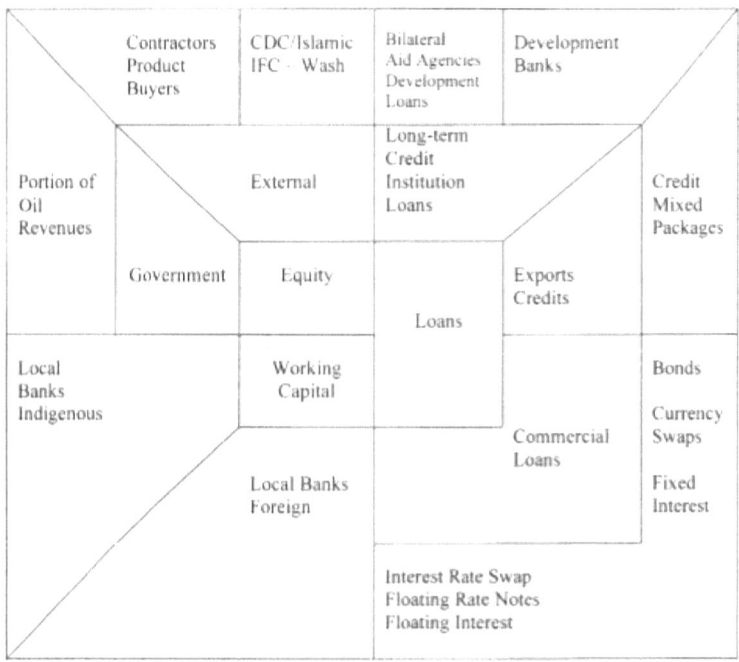

Figure: Sources of finance, adapted from (Smith, 2010)

5.3 Project Finance

Project finance can be defined as a phenomenon of financing a specific legal entity whose revenues and cash flows will be received by the lender as a financial source which the loan will be repaid (Smith, 2010)

This implies that the contracts, economies, cash flows and the assets of the projects are separated from the sponsors to enable a limited recourse with the aim of lenders assuming some of the project's risk of commercial success or failure.

To be precise, (Nevitt, 1983) defined project finance as "the financing of a particular economic unit in which the lender is satisfied to look initially at the cash flows and earnings of that unit as the source of funds from which the loan will be repaid and to the assets of the economic unit collateral for the loan''

It is worth noting that if the revenues for the project is unable to fund the debt service fully, the lenders have no claim against the owner beyond the assets of the project; the project, in effect, being self-liquidating and self-funding regarding financing. Hence project finance provides no recourse.

Projects financed privately are largely funded by the combination of equity and debt capital, unlike customary public sector projects that have their capital costs largely funded by loans raised by the government. The ratio between these two types of capital differs between projects.

One of the most difficult operation is financing a project especially with the highest risks taking place at the construction phase. Such challenge is dealt with by providing equity finance before revenue is generated by the project. Also, lending organizations must ensure that risks on loans provided to sponsors are covered by guarantees.

Factors affecting private finance package include loan currencies, loan schedule and the possible effect of associated risks. To adequately maximize private finance for obtaining public sector project infrastructures, it is fundamental that the finance is considered individually.

The government must ensure that necessary support is given to projects that will be commercially viable in the private sector to make sure that the economic, social benefits or projects value be utilized by the end users.

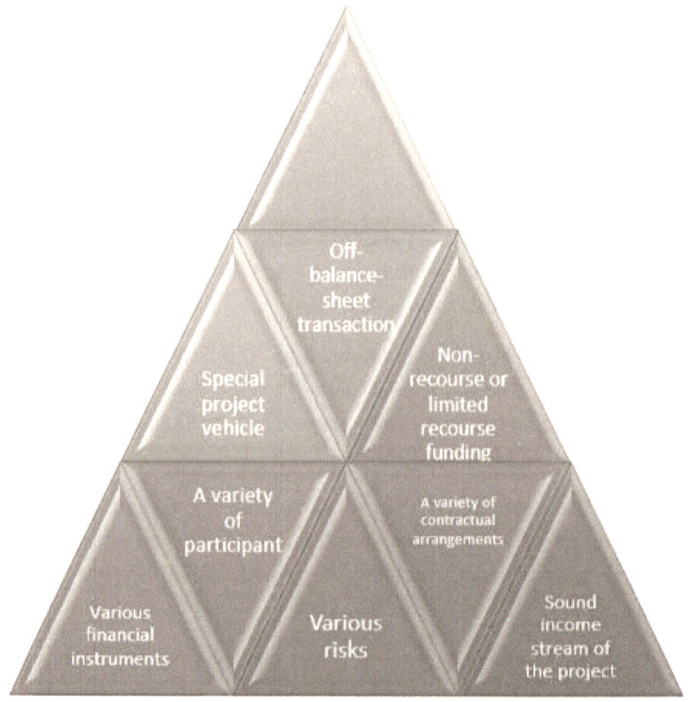

Figure: featuring project finance basic (Smith, 2010)

5.4 Financial Instruments

It is obvious that projects do not progress without finances as all projects require financing. Nevertheless, there is often variation in amount and nature of the required finance as the project changes from one phase to another.

The expenditure rate for many projects experiences drastic change as the project progresses from the appraisal stage which includes analytical skills and human expertise to the design stage, which then leads to assembly and construction stage and lastly to the operational phase.

The three major stages a project passes through is

1. Project Appraisal
2. Project construction/implementation
3. Project operation

A variety of factors determine the cash flow curve for individual projects including obtaining statutory approvals; time is taken to set up project objectives, finalizing designs, contract agreements, amount and rate of construction, finalizing financial arrangement and speed of project operation as well.

A classic project with a negative cash flow will require financing from outside until it breaks even. The curve shape at the initial phase of the project displays a need for a relatively less financing. However, as the project advances to the construction phase, a sudden financial increase is required which peaks at the final stage.

The rate of spending is also revealed by the steepness of the curve as steeper curve shows a higher financial need. Once revenue generation commences after project commissioning, a financial requirement from an external source is lessened until there is sufficient revenue generation to operate and maintain the project effectively.

However, after the break-even point, the project may need finance for interim periods to cover up for the mismatch between receipts and payments.

The factor for raising required resources for project investment in project financing is determined by the future cash flow. Hence, it is the responsibility of the project finance team to ensure that the cash flow is adequately prepared to meet such project needs and attract the interest of potential investors.

Therefore, the project manager must be empowered with adequate financial instruments and financial markets knowledge to trade and muster all requirements to finance the project effectively. A channel of generating finance for projects is through their investment activities which are via selling/issuing securities.

These securities are popularly known as financial instruments which serve as a form of claim on project cash flow in the future. Simultaneously, these financial instruments have a liable claim on the project asset, hence serving a security measure should the expected future cash flow fails.

The claims these instruments have on the cash flow and assets of the project vary regarding nature and seniority. Other financial instruments are called "mezzanine finance" since they share the features of both debt and equity.

Securities issued by the project which is subjected to the payment of a specified amount at a time is known as *Debt instruments*. It consists of term loans, debentures, export credit, supplier's credit and buyer's credit. It is senior to all other claims on the project cash flow and assets.

In the same vein, *Ordinary Equity* refers to the ownership interest of common stockholders in the project. On the balance sheet, equity is represented as *total assets less all liabilities*. It is lowest of the rank hence, the last claim on the cash flow and assets of the project. Equity is usually represented as ordinary shares and preference shares.

Mezzanine finance occupies an intermediate or midway position between the senior debt and common equity. This kind of finance takes the form of subordinated debt, junior subordinated debt, and preferred stock, or some combination of each.

Bonds such as plain vanilla bonds, junk bonds, Euro bonds, income bonds, discount bonds, floating rate bonds, warrants and convertible bonds, revolving underwriting facility (RUF), and note issuance facility (NIF) are all Mezzanine finance instruments.

Other instruments like venture capital, leasing, aid and depository receipts can be utilized by a project.

5.5 Financial Engineering

Specialized financial instruments are utilized by financial engineers to improve financial performance like engineers using special tools and instruments to accomplish engineering perfection.

The creative application and development of financial technology to solving financial challenges and exploit financial opportunities is known as *financial engineering*. It can also be defined as the use of financial instruments to restructure existing financial profile to having more desirable characteristics (Smith, 2010)

Techniques such as the development of derivative instruments and securities, hedging, and financial risk management, forecasting financial markets, modelling, asset allocation and investment management, and asset /liability management are rapidly applying financial engineering.

5.6 Debt Financing Contracts

A standard financial loan package should accomplish the following fundamental objectives.

1. Maximize long term debt: This ensures that the project unit incurs its debt, hence not affecting the balance sheet of the sponsor's parent company.

2. Maximize fixed-rate financing: Projects risks are reduced using long term export credit facilities or subordinated loans with low-interest rates.

3. Minimize refinancing risk: Projects could acquire additional problems because of cost overruns, thus standby credit facilities from lenders and additional capital from supporters should be obtainable.

The contract between lender and sponsor can only be determined when the lender has adequately assessed the viability of the project by sufficient information. It is common that the lender looks to the project as a source of repayments instead of assets in construction projects.

It is therefore beneficial that the characteristics below be considered by the lenders to take the right decision

1. The total size of project: This is a key factor in determining the amount of cash flow needed and the required effort to raise the capital, including the internal rate of return on the project and the equity.

2. Break-even dates: These are dates when investors see a return on their investments.

3. Milestones: Important dates related to the financing of the project

4. Loan summary: the true cost of each loan, the amount drawn and the year in which drawdowns reach their maximum.

A term sheet is often prepared by the lender to manage different types of risks. This term sheet, depending on the agreement among the project teams, outlines the obligations and rights of lenders, it also describes default conditions and remedies, and serve as the bid document for accessing capital markets.

The contents of the term sheet may include a bank loan, index or coupon bond, drawdown of loans, equity, dividends, preferred shares, interest during construction and operations. Others are lenders fees, interest rate risk (IRR), net present value (NPV), the coverage ratio, payback period of loans, a standby facility, the working capital and the debt service ratio.

These components aid to determine whether a loan is authorized for a project, and provide a mechanism to evaluate the financial parameters of a bid.

It is expedient that lenders are satisfied that the project is viable for each risk analysis. The following criteria help to access the projected cash flow.

1. The debt or service ratio: Annual cash flow available for debt service divided by the debt service

2. The coverage ratio: NPV of future after-tax cash flows over either the project life or the loan life divided by the outstanding loan balance

The coverage ratio is a criterion for evaluating projects specific to oil and gas whereas debt or service ratio is utilized to evaluate limited resource financing in all types of industry.

For instance, the debt will be repaid with no margin of error in cash-flow projections for a project with a 1:1 coverage ratio. Ratios less than one infers to a no repayment debt over the term of the calculation and ratios more than one gives a measure of relief should there be variations by assumptions.

The term sheet may be used to express such ratios for one or a few risk analyses and in the event

Where the ratio is less than the requirement of the lender, then the loan package may be considered impracticable commercially for the risks to be covered.

5.7 Appraisal and Validity of financing projects

Project sponsors should examine the following to assess the attractiveness of a financial package adequately.

1. Interest rate, debt/equity ratio including percentage being financed

2. Repayment period, the currency of payment, legal, management and syndication fees, guarantees fees and documentation of all kinds including draw down loans, activation, and application.

The fundamental financial criteria needed to be considered in projects are:

1. The cost-effectiveness of finance, i.e. effective utilization of finance at fixed rates to lower risks and source for long-term finance

2. The project must have clear and defined revenues enough to service principal and interest payments to provide commensurate equity return.

The strength of the financial security package, the perception of the country's risks and limits and a full understanding of local capital markets are aspects to be considered in the selection process of capital sources.

Full financing of projects can be entirely carried out from debt. However, the lenders take the risk that revenues to be generated will be sufficient to pay off the debt by the end of the concession period.

One major determinant factor to be considered in project finance is the ability to provide security to non-recourse or limited-recourse lenders. For instance, if a sponsor defaults using a non-recourse finance package under a project, the lender may be left with a partly completed facility with almost no market value.

The security devices below can help protect lenders:

1. Revenues received in one or more escrow accounts is managed by an independent escrow agent different from the sponsor company

2. Various contracts benefit entered into by the sponsor is assigned to a trustee for the benefit of the lender including construction contracts, performance bonds, supplier warranties and insurance proceeds.

3. In cases of technical or financial default before bankruptcy, lenders may insist on the right to handle the projects and bring in new contractors, operators or suppliers to finish the project.

4. Export credit agencies and lenders may beckon on government support such as standby subordinated loan facilities which are functionally almost equivalent to sovereign guarantees.

In conclusion, funding projects successfully should include limited and non-recourse credit, debt financing completely in local currency, equity finance in relatively strong currencies, major innovations in project financing, project creditors who are confident, and government which accept some project risks and provide limited resources.

5.8 Risks

The early stages of the project appraisal are the best time to identify possible associated risk with the project before it undergoes analysis and allocation. It is apparent that investors and lenders will only be enticed to projects that guarantee suitable returns on the capital invested.

Some risks associated with projects include those within the control of a few individuals related to the project. Other risks may not be within the control of any reasonable parties but may be insurable at a cost while the last kind of risks is the uninsurable ones.

A more accurate estimate of the duration, final costs and revenues of a project can be determined by adequately identifying possible risks at the evaluation stage of a project.

5.9 Financial Risks

Financial risks may comprise of off-take agreements, take or pay terms, foreign exchange risks, debt service risk, and effect of escalation. The mechanics of providing adequate working capitals, raising and delivering Finance is known as financial risk.

Financial risk may be apparent during the operation phase when machinery is running to specification but does not provide adequate revenue to cover operation costs and debt service.

PILDEC could be an acronym for financial risk

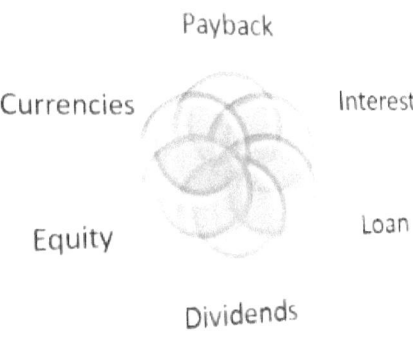

Figure showing financial risks element

Payback: This consist of fixed payments, loan period, cash-flow milestones, discount rates, the rate of return, payments scheduling, etc.

1. Interest: It includes changes in interest rate, existing rates, type of rate (floating or capped, fixed)
2. Loan: It comprises of availability of loan, servicing cost loan, default by the lender, type, and source of loan, standby loan facility, debt/equity ratio, holding period, existing debt, covenants.
3. Dividends: time and amounts of dividend payments
4. Equity: Take-up of shares, type of equity offered, institutional support and so on
5. Currencies: mixed currencies, ratio of local/base currencies, currencies of loan

Revenue risks

Revenue generation risks are usually considered by meeting demands and could include

1. Demand: This deals with demand over the concession period, the demand associated with existing facilities, the accuracy of demand and growth data and capacity to meet the increase in demand.
2. Toll: This consist of shadow tolls, toll level, revenue currencies, market-led or contract-led revenues, tariff variation formula, regulated tolls, take and pay payments.
3. Developments: This involves changes in streams of revenue from associated developments

Commercial Risks

Commercial risks help determine the commercial practicability of a project. Market and revenue streams are affected by commercial risks.

This kind of risks can be classified under six major topics

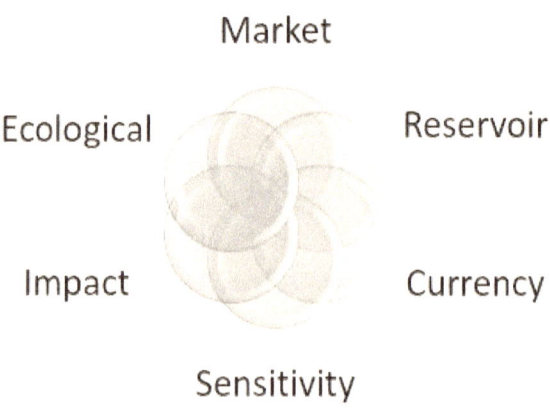

Figure showing the main commercial risks elements

Market: This covers demand changes for a product or facility

1. Reservoir: It deals with changes in input source
2. Currency: Covers fluctuation in exchange rates, devaluation, convertibility of revenue currencies
3. Sensitivity: Project location, existing environmental constraints, imminent environmental changes
4. Impact: effect of environmental impact, changes in environmental consent, effect of environmental impact
5. Ecological: ecological changes during concession period

Project finance analysis can be summarized under these four key titles

1. Cost analysis: This establishes a minimum project cost including development, construction and operating costs
2. Financial market analysis: This comprises of cost and conditions of project financing including availability data.
3. Market analysis: This assesses the commercial practicability of the project via demand forecasting and establishes a maximum price.
4. Financial analysis: It establishes the relationship between revenues and costs by juxtaposing the cost, the market, and financial market analysis together.

Decreasing risks in international investment analysis can be accomplished by the following methods

1. Reduce the minimum payback period
2. Ensure that the required rate of return of the project investment is raised
3. Adjust cash flows to reflect the specific impact of a risk
4. Adjust cash flows for the cost of risk reduction

Risks Allocation

Lenders tend to minimize projects risks by engaging the following responses

1. Completion risk: This consist of firm date, cover by a fixed price, turnkey construction with stipulated liquidated damages

2. Cash-flow risk: Here, escrow arrangements are used to cover forward debt service and protect against possible disruption, and take out commercial insurance

3. Inflation and foreign exchange risk: cover by government ensures tariff adjustment formula, minimum revenue agreements and assures convertibility at agreed exchange rates

4. Insurable risk: ensures policy to cover cash flow shortfalls especially at the pre-completion phase of the project

5. Performance and operating risk: Here, export credit agencies or multilateral investment agencies cover political risk insurance
6. Commercial risk: Export Credit Guarantee Department (ECGD) covers such insurance policies

QUESTIONS
CHAPTER 5: PROJECT FINANCE

1. Project type, revenue generation and location often determine cost of finance and its components
   a) True b) False c) No idea

2. If the revenues for the project is unable to fully fund the debt service, the lenders have no claim against the owner beyond the assets of the project
   a) True b) False c) No idea

3. ....is the first stage a project passes through
   a) Project implementation b) Project Appraisal c) Project operation

4. Securities issued by project which is subjected to payment of a specified amount at particular time is known as *Debt instruments*
   a) True b) False c) No idea

5. A type of risk in project finance is called ....
   a) Engineering risk b) ecological risk c) revenue risk

ANSWERS
1. A   2. A   3. B   4. A   5. C

CHAPTER 6

PROJECT PLAN WITH TIME MANAGEMENT

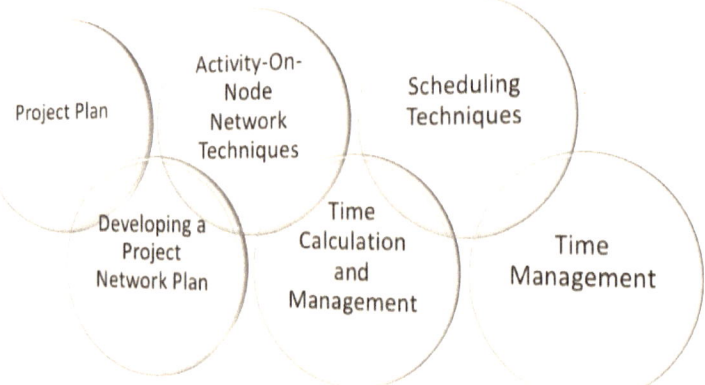

6.0 PROJECT PLAN WITH TIME MANAGEMENT

So, in this chapter, we will be discussing time management.

1. Project Plan
2. Developing a Project Network Plan
3. Activity-On-Node Network Techniques
4. Time Calculations and Management

6.1 Project Plan

Project execution is the implementation of a task under the constants of time pressure, so time is a critical measurement for project effectiveness. The essentiality to having a project plan is due to every one of these difficulties in executing projects.

A project plan provides necessary information for each team member whose work is identified with the project. Usually, the project plan is a simple planning tool at the commencement of the project. However, it becomes one of the most critical control instruments during project execution and after completion of the project.

It is also an estimation of whether the project has achieved its objective. This chapter will disclose how to develop a project plan and current practices of project plan approach.

The distinctive ways to approach project networks will be described: (Passenheim, 2009)

| Pert | Critical Path Method | Activity on Arrow | Activity on Node Network | Gantt Charts. |

These distinctive methods and methodologies have been picked because they all are utilized as a part of projects, and many firms have characterized one method as a standard within their firm

6.2 Developing a Project Network Plan

A project network is a basis for scheduling a budget plan, machines, equipment, meetings, correspondences, the estimated time utilization and the beginning and the completion dates.

The following terminology has been defined to structure a project network and set a standard which many individuals can utilize.

An **activity** is a component of the project network; activities are defined duties aimed at meeting project objectives. An activity needs time to be finished. It consists of resources like budget, personnel, space and, in many occurrences, relationships (Passenheim, 2009).

An activity in a project network indicating which duties must be performed for us to progress, which assets are required and what number of them. As the name infers, **a parallel activity** is a task which is executed at the same time, parallel to other tasks. When carrying out parallel activities, there will be an increase in the utilization of resources due to their concurrent demands.

A **merge activity** is one activity which takes after on more tasks; parallel tasks meet up in this activity. A succeeding activity must only commence when all prior processes are completed.

An **event** is something which does not utilize the project time; an event is often a date. It is characterized by beginning or ending a date when something is executed or delivered. Likewise, it could be the commencement or the ending and many dates between.

A **path** is a link between the dependent tasks or activities. It has no time interval and is the visualizing of the interconnectivity between events. **The critical path is the shortest duration of the project.**

The most important activities are placed on the critical path due to its relevance to the project, if an activity on the critical path is delayed, it will have a high consequence which will alter the total project duration.

The **critical path** is presumably the most vital result after drawing a project network plan. Each project has such a path and the workflow of all critical tasks add entirety up to the critical (shorted) duration time.

Other tasks which are not on the critical path do not constitute a major problem to project completion. However, they could result in project delay which will not effect on the completing the project. The project team can effectively act and react to situations during project implementation with the knowledge of critical path.

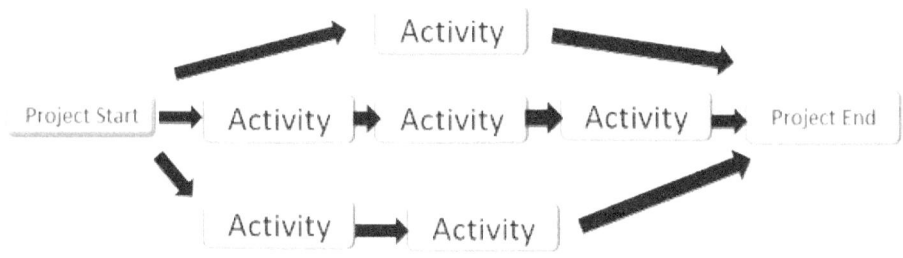

Example of a Project Network Plan

The fundamental rules below should be implemented during the development of a project network plan:

Networks flow from left to right. That is an absolute necessity as this approach guarantees that the project group knows how to comprehend the network.

An activity must be utilized once: it is not permitted to make circles, where an activity must be executed several times. On the occasion that a similar task must be repeated, a new activity must be drawn up. An activity cannot commence until all prior activities have ended.

This is quite clear as it is unrealistic to start something until the first preceding tasks have been completed. This reality must be displayed in the network.

Arrows on networks display the flow of precedence; they can cross. To demonstrate the flow of procedure and to show which activities must be done, arrows are utilized.

Every activity must have a peculiar identification number: numbers help to identify network orientation in the network easily. The number of the activity reveals the flow of work. With numbering of activities, it is much easier to follow through a project path.

The numbering of activities ought to be done in ascending order, which implies the start activity ought to have the lowest number and the last activity should have the highest number.

Every activity needs a unique identification code; most computer programs accept numeric and alphabetic codes or a mix of the two. The project manager ought to leave gaps between numbers (5, 10, 15,), so he can include other activities later. Most times, activities skip the mind of the project manager and could be quite difficult to draw a perfect project network from the start. Thus, they must be broken down into smaller activities.

Clear tasks must be utilized to display the starting and the ending of the project. The project network is a plan which needs to support the work process in the project. Everyone needs to understand his/her specific responsibility in the project clearly.

To show the starting date is critical because it demonstrates to the project team that they need to meet the objective of the project with the beginning of another activity. It is likewise critical to define the end date.

The outlining of the starting and ending tasks is connected to the project and the definition of dates, which need to meet while the project is progressing.

Milestones are activities characterized by a distinctive task in the project. They are control points of the work carried out up to that date. The milestones give the project team and particularly the project manager the capacity to quantify the work process of the project while implementing it.

This helps them to monitor and control the project while in the process, allowing for quick response and reaction to adjust or do something in the preceding project time to achieve project objective missing the deadline. Milestones are essential to control tools in utilizing project networks.

"If-than-else" or conditional statements are not to be utilized as a part of project networks. This rule is more of a mental or psychological one; conditional statements can reflect conceivable uncertainty and the reader of the project network may get the feeling that the activity is not manageable.

To demonstrate that the project is efficient and will be successful it is imperative to use statements which reveal that the activity will be successful.

The work breakdown structure is the first phase in setting up a project network.

The second step is to improve the activities with adequate information.

The necessary resources required to complete an activity are the most vital: labour, equipment, time, costs, space and so forth. It is also vital to find out what the interconnectivity and the dependencies of the activities are.

The third step in developing the project is the execution of the information into the project network plan.

6.3 Activity-On-Node Network Techniques

Increase in technology in recent times have been accompanied by figures and illustrations through the high-quality graphical output of personal computers leading to the expansion of the activity-on-node (AON) network plan.

An activity is characterized by a node which can take diverse forms yet is regularly represented to as a rectangular box. The activities are linked by arrows between these boxes. The arrows represent dependencies between the diverse activities and the specific order in which the estimated tasks must be implemented.

The length, width or slope provides no information about operational hours, workload and so on., they simply enhance the readability of AON-plan by enhancing the visual lucidity of the dependencies between the activities of a project.

In establishing activities into a project the project manager and his/her team members need to characterize the relationships of every activity in the project context. This can be carried out by providing answers to the following three questions for each single task:

Which activities must be completed before commencing this one?	Which activities can directly begin after completing this one?	Which activities can be implemented in parallel to this one?

The answers to these questions give the project team the required knowledge on the relationships for each task either the predecessor, successor and parallel relationship between them. Acquiring this information is critical for the development of a graphic flow chart which enhances visibility of dependencies and relationships between different activities

Figure: Example 1 of AON

The figure above shows a standard network plan. All the tasks are carried out in succession, and the project manager is fully aware that activity B cannot commence until A is done, and activity C has to hold on until task B has been implemented.

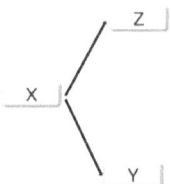

Figure: Example 2 AON

The figure above reveals that the activities Y and Z must wait until task X is completed.

It also shows that activities Y and Z can be done in parallel.

The graph demonstrates that the two activities can be executed concurrently to save time, for instance. The project manager makes the final decision on how the activities are organized.

He/ She i.e. the project manager organizes this based on the resources available by organizing activities in a row form as opposed to working in parallel.

A project manager also must frequently manage limitations on the availability of construction workers to organize activities for example, an office construction project. He needs to choose whether, e.g. the foundation for the office structure or the carport is to be done first, although these tasks could be executed concurrently.

The number of arrows running out of a node in an AON-Plan reveals some activities to be done after. The activity X in the figure above is called burst activity since more than one bursts from its node.

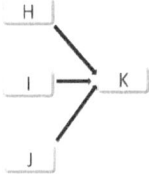

Figure: Example 3 AON

The figure reveals a possible alternative situation in an AON-network. Task H, I and J can be executed parallel or concurrently if the required resources are available and possible constraints are absent. Activity K has to wait for task H, I and J to be completed until it can be started. For example, a painter cannot start his work for a house until the foundations, and the walls are ready.

In this case, activity K is called a merge activity since multiple tasks must come to an end before it can commence. In figure below, activities H and J can be done in parallel. Also, task K and task I can take place concurrently but have to wait until the preceding activities H and J have been done.

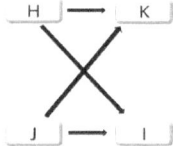

Figure: Example 3 AON

With this fundamental knowledge, it is important to follow the next example in developing an AON-network. It is also vital to recall the essential rules highlighted earlier.

Arrows can cross over each other like in figure second above; as mentioned earlier, the lengths of the arrows relates no information on the duration of the task. It is important to develop an accurate and logical inclusion of all project activities including all their time estimations and dependencies.

The table below contains information about simplified activities that have to be done to install a new suspension bridge.

ACTIVITY	DESCRIPTION	PRECEDING ACTIVITY
A	Approval application	None
B	Fundamental Installation	A
C	Fabricate steel elements	A
D	Fabricate tower elements	A
E	Fabricate steel ropes	A
F	Fabricate supporting elements	A
G	Trasport items to building site	C,D,E,F
H	Suspension bridge erected	B, G
I	Fine tuning	H
J	Testing	I

Table: Activity list, adapted from (Passenheim, 2009)

The first steps in developing an AON network based on the information stated above in Table above are shown in the figure below. Activity A is the first node which is drawn since it has no preceding activity. Activities B, C, D, E, and F are directly reliant upon activities A.

The project team members have to hold on until they receive the required approval for building such a suspension bridge before they commence the installation of the foundation and fabrication of all needed units. The preceding activities are connected by an arrow with task A since the other succeeding activities can be done simultaneously when task A is "ready". The figure below shows the complete network of our example project. The project manager can find all activities and their dependencies in a graphical way.

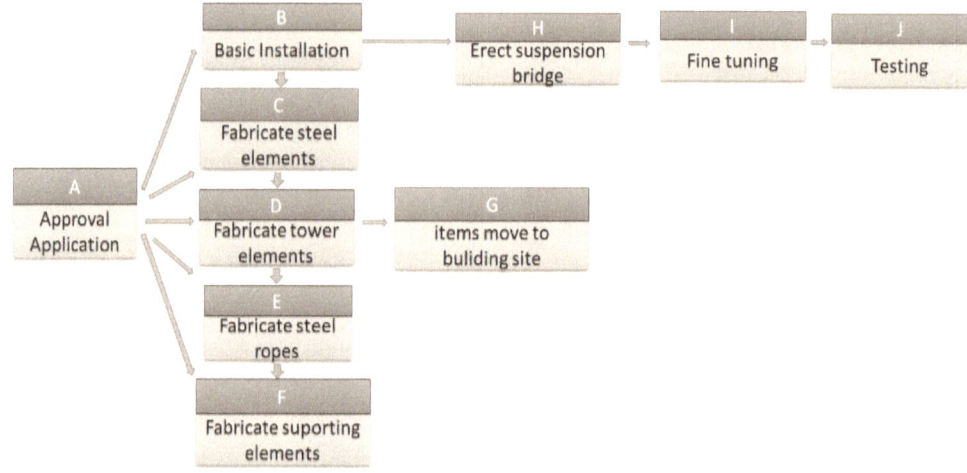

Figure: AON Example Suspension Bridge

6.4 Time Calculation and Management

Time Calculations

The calculation of the start and finish times of the activities is the first step to developing an AON-network. The estimated times for each activity should have a realistic base to create a reliable AON network.

The estimated time for the suspension bridge example (Table above) are simplified and gives a fundamental understanding of AON-networks calculations. The project manager can easily carry out simple computations develop the **forward** pass – earliest times and **backward pass** – latest times.

The **forward pass** questions to be answered are:

I. How rapidly can the activity commence? (Early Start – ES)
II. How soon can the activity be completed? (Early Finish – EF)
III. How rapidly can the project be completed? (Expected Time – ET)

The **backward pass** questions to be answered are:

I. How late can the activity commence? (late start – LS)
II. How late can the activity be completed? (late finish – LF)
III. Which activities represent the critical path (CP)? (This is the longest path in the network which, when delayed, will delay the project.)
IV. How long can the activity be delayed? (**slack or float** – SL)

4.4.1 Forward Pass – Earliest Times

The forward pass starts or commences with the first project activity and spots each path (chain of sequential activities) through the network until the last project activity is reached. As the path progresses, the activity times are added. **The longest path indicates the project completion time for the plan and is known as the critical path.**

Table 4-2 provides information about the duration of each activity that must be done to complete the wind energy plant example project. (Passenheim, 2009)

ACTIVITY	DESCRIPTION	PRECEDING ACTIVITY	DURATION (DAYS)
A	Approval application	None	5
B	Fundamental installation	A	10
C	Fabricate steel elements	A	10
D	Fabricate tower elements	A	20
E	Fabricate steel ropes	A	15
F	Fabricate supporting elements	A	10
G	Trasport items to building site	C, D, E, F	5
H	Suspension bridge erected	B, G	10
I	Fine tuning	H	5
J	Testing	I	5

Table: Time Estimation

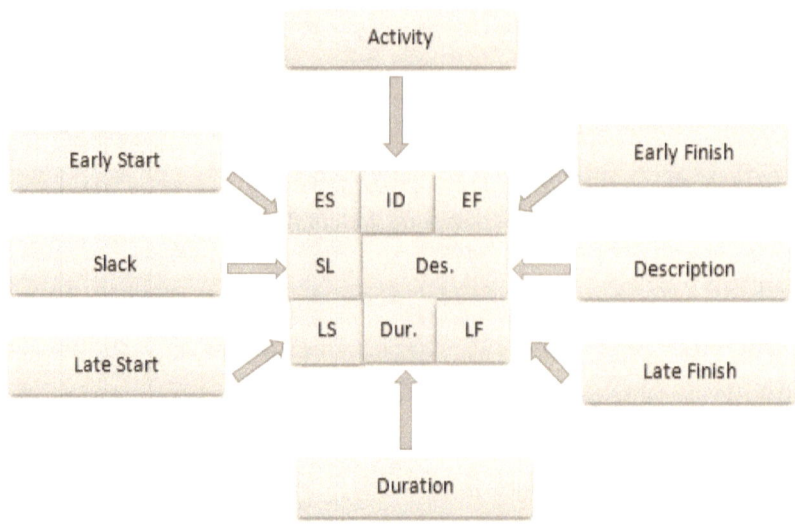

The figure above: shows an enhanced node layout where additional information can be stored. (Node with extended information)

For the suspension bridge example, the times for each activity are taken from the list as shown in Table 4-2 and were put into the field "duration" (Dur) for each activity. For example, activity A has a duration time of 5 days and activity H of 10 days

To start with the calculation for the forward pass, a start time should be defined. In this case, the start time is 0. In "real life or practice" developing a project network is dependent upon calendar dates with holidays, weekends and so on. The suspension bridge example is simplified to support the placement of the AON- networks principles.

The early start (ES) for the first activity, A, is zero. This time is in the upper left corner of the activity A-node in the figure above. The early finish (EF) for activity A is 5,

Since (ES + Dur = EF; 0 + 5 = 5).

Activity A is the predecessor to activities B, C, D, E, F. The earliest start date for these activities is 5 because all of them are directly following on from activity A and hence they have to wait until activity A is finished. To compute the early finish (EF) for activity B, C, D, E, and F the formula ES + Dur = EF is used.

1. EF(B) = 5 + 10 = 15
2. EF(C) = 5 + 10 = 15
3. EF(D) = 5 + 20 = 25
4. EF(E) = 5 + 15 = 20
5. EF(F) = 5 + 10 = 15

It is convenient to select the right earliest start for activity G since activities C, D, E and F are preceding this activity. Although, there are three possible answers only one is correct. The calculations of the early finish of activity C, D, E and F produced three different results: 15, 20 and 25.

This is because all activities immediately preceding activity G must be completed before G can begin, the only possible choice is 25 days. It will take activity D the longest duration to complete. It also controls the early start of activity G. Activities B and G are preceding the next, activity H.

The largest early finish of both leads to the right solution for the early start date of activity H. As initially calculated, the early finish of B is 15, and the early finish of G is 30, so the early start of activity H is 30.

The next activity I have only one preceding activity. After calculating the early finish of H by using following formula:

ES + Dur = EF (30 + 10 = 40)

It can be carried to me, where it becomes its early start. The same procedure is used to compute the early start for the last activity J. So the EF of activity I (40 + 5 = 45) becomes the early start of I.

In the same vein, the early finish of J (45 + 5 = 50) shows the earliest possible time the whole project can be completed under standard conditions.

Rules for forwarding pass computation:

I. Activity times along each path in the network (ES + Dur = EF) are added
II. The early finish (EF) is carried to the next activity where it becomes its early start (ES), unless
III. The next succeeding activity is a merge activity. In this case, the largest early finish number (EF) of all its immediate predecessor activities is selected

When the forward pass is calculated, the suspension bridge example displays the graph below

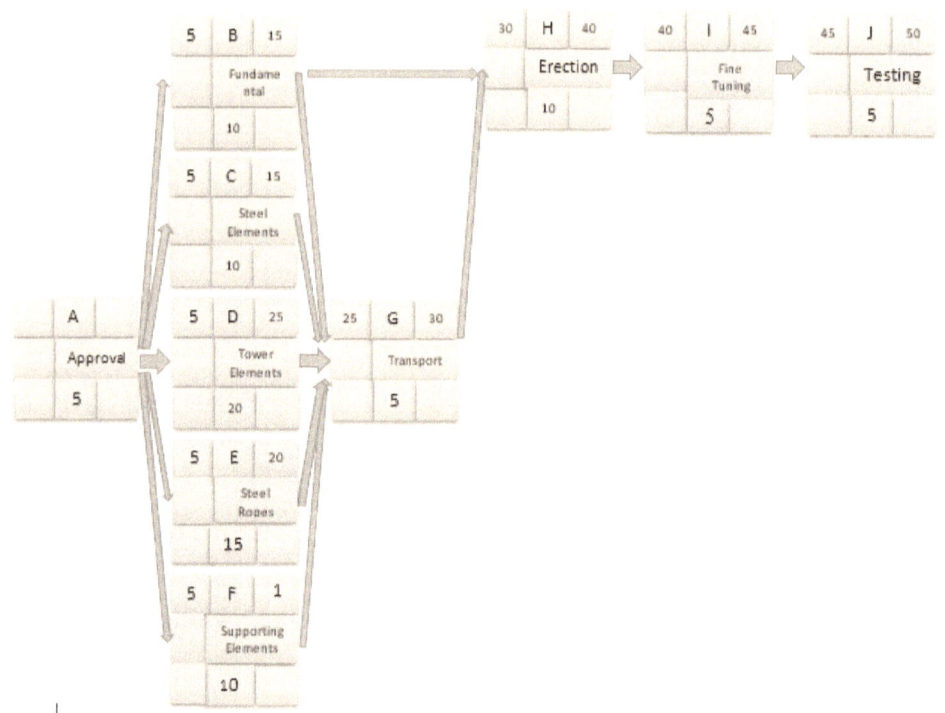

Figure above: Forward Pass adapted from, (Passenheim, 2009)

Backward Pass – Latest Times

The *backward pass* calculation starts with the last project activity on the network. Each path is traced backwards and activity times are subtracted to find the (LS) and finish times (LF) for each activity. The late finish, for the last project activity, must be selected before the backward pass can be computed.

In the early stages of planning, this time is usually set equal to the early finish (EF) of the last project activity (or in the case of multiple finish activities, the activity with the largest EF). In situations which include cases of an imposed project duration with a deadline, and this date will be used.

In the suspension bridge example the 50 days of early finish of the whole project are also accepted like the latest finish of the project, and therefore the EF of activity J is carried to its LF. To compute the backward pass, three basic rules are needed for analogue to the forward pass.

I. Activity times are subtracted along each path starting with the project end activity (LF – Dur = LS)

II. The LS is taken to the next preceding activity to establish its LF, unless

III. The next preceding activity is a burst activity; should the smallest LS of all its immediate successor activities be selected to establish its LF

These rules are used to calculate the backward pass of the wind energy example. First, the LF of activity J (50 work days as stated above) is subtracted with its duration (LF – Dur = LS; 50 – 5 = 45). The calculated LS of J is taken or carried directly to activity I where it becomes its LF.

The LS of activity I (45 – 5 = 40) is again directly transferred to activity H (LF). The LS of activity H (40 – 10 = 30) may directly affect the two activities G and B. In the case of activity G, the LS of activity H is directly transferred to the LF of G since activity H is it's only immediately the following activity.

The LF of activity B is controlled by the LS of activity H. The latest activity B can be finished in 30 days. The LFs of activities C, D, E and F depend solely on activity G, therefore, giving an LF value of 25. The LS dates of activities B, C, D, E and F which affects the LF of the first activity A are computed below.

1. LS(B) = 20 – 10 = 20
2. LS(C) = 25 – 10 = 15
3. LS(D) = 25 – 20 = 5
4. LS(E) = 25 – 15 = 10
5. LS(F) = 25 – 10 = 15

As with the rules outlined above, the smallest LS of activity B, C, D, E, and F serves as the right choice for the LF of activity A, in this case, the LS of D, since activity D requires the longest time to complete in this comparison. The LS of activity A (5 – 5 = 0) completes this backward pass, all latest activity times are known.

Figure below: Graph after calculations are made, adapted from (Passenheim, 2009)

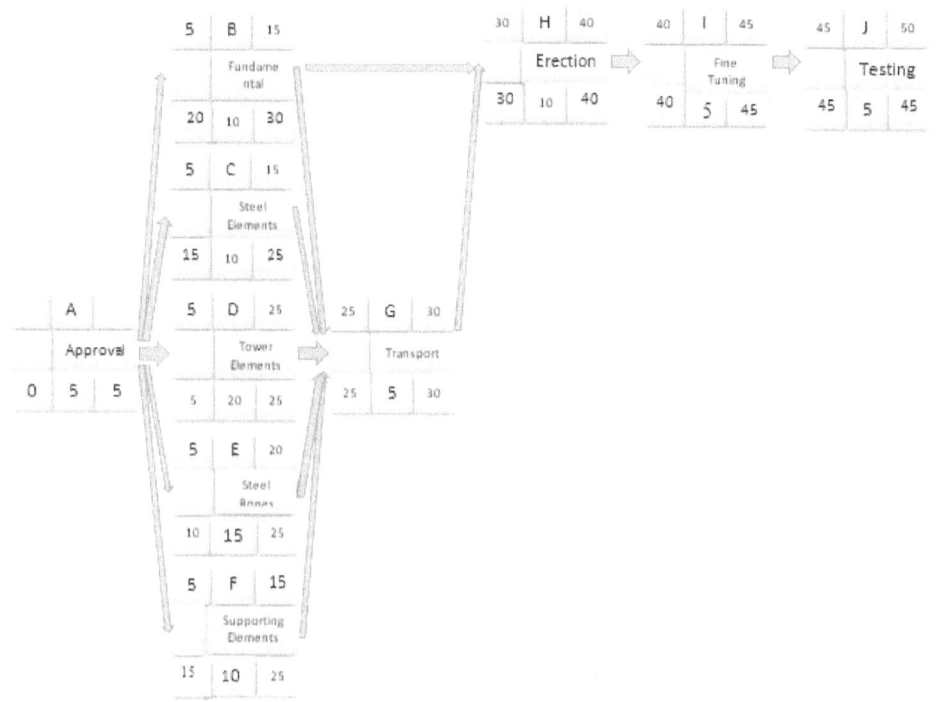

Figure above: Finished Graph

Identification of Float or Slack

Activities that could be delayed can be determined by calculating "slack" or "float" when the forward and backward passes have been computed. Total slack or float for an activity is simply the difference between the LS and ES (LS – ES = SL) or between LF and EF (LF – EF = SL).

Figure **(all d graph)** 2-11 shows different examples, slack for activity B is 15 days, for activity E 5 days and 10 days. The slack displays information of the amount of time an activity can be delayed without delaying the whole project. If slack of one activity in a path is utilized, the ES for all activities that follow in the chain will be delayed and their slack reduced.

The use of total slack must be coordinated with all team members in the activities that follow the chain. After slack for each activity is computed, the critical path(s) is (are) easily identified.

When the LF = EF for the end project activity, the critical path can be identified as those activities that have **LF = EF** or a slack of zero (LF – EF=0 or LS – ES = 0). The critical path in the suspension bridge example is represented by activities A, D, G, H, I and J

(Figure (above, finished graph). A delay in one or more of these activities would delay the whole project.

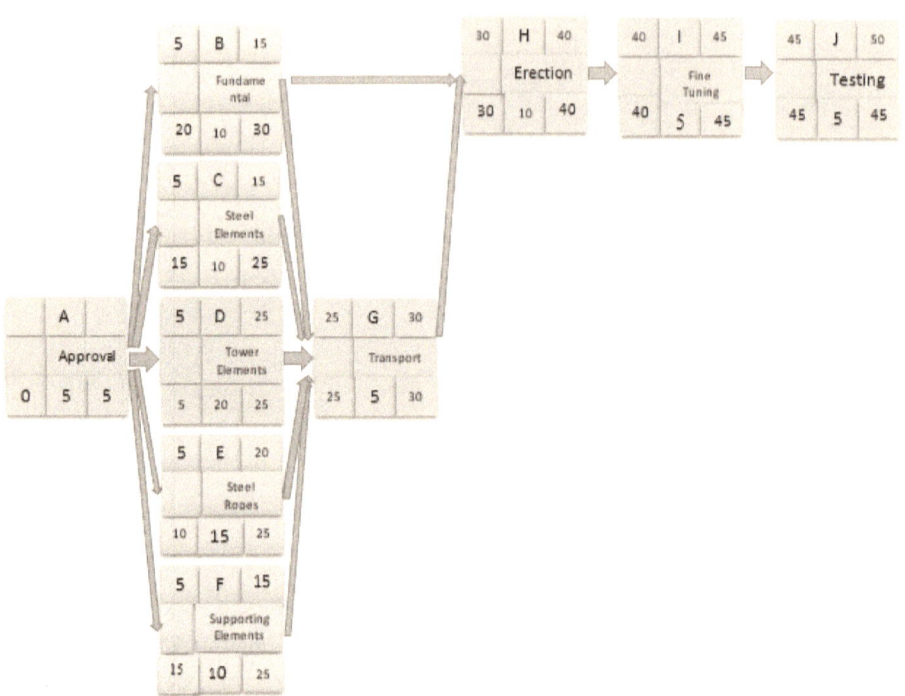

Figure above: Critical Path

A network schedule with a single critical path and non-critical activities that enjoy significant slack would be named insensitive. On the other hand, a sensor network would be one or more critical paths and/or non-critical activities with almost no slack.

Under these conditions, the first critical path is significantly more prone to change once work gets going on the project.

AON-networks have multiple benefits. The coherent structure, ease of drawing and the basic components can be easily utilized to assist the project manager with designing such networks. Another benefit includes the involvement of team members indirectly, for example, first-level managers can easily comprehend the significant points and issues of a project rapidly by utilizing an AON-Network because the graphical representation portrays reveals activities that impact the project into structured context.

The calculations expected to build up an AON-Network are very straightforward and simple to handle. However, an AON-Network without a graphical representation is futile, since only a table including all figures and facts with a graphical layout can be understood proficiently.

AON-Networks made with recent computer programs can assist the project manager to effectively structure his day by day work in a proficient path and to focus on other significant activities.

6.5 Scheduling Techniques

Network scheduling techniques shape the basis for all planning and forecasting strategies. It also enables management to effectively administer and utilize its assets and resources to accomplish time and cost objectives. Managers can adapt t7.7o the complexities, masses of information and tight deadlines that are associated with the highly competitive organization by making use of these techniques.

These techniques give room for more transparency in project management and help to easily identify dependencies between activities, to schedule risks, to distinguish critical paths and to evaluate the impact of delays on project completion.

There are diverse scheduling techniques, yet the most widely recognized ones are network charts like the AON strategy and the Program Evaluation and Review Technique (PERT).

The Program Evaluation and Review Technique is a network analysis technique which utilizes the AOA or AON approach to estimate duration of the project. PERT can manage activities uncertainty during completion times. It also helps to develop better schedules to maximize cost and time resources.

This serves as a great benefit in comparison to the critical path method. The CPM utilizes fixed time estimates for specific activities and more deterministic.

A three-point estimate for each activity is required for the performing of PERT appraisals. However, time varieties that could highly affect the finish time of a complex project are neglected.

A three-point estimate is the duration estimate of an activity which consists of

An Optimistic estimate

Most Likely estimate

Pessimistic estimate.

1. The optimistic estimate depends on an ideal situation, i.e. a best-case scenario. For the most part, it is the shortest time complete the activity.
2. The Most Likely estimate is based on an expected scenario or situation. Its completion time has the highest probability.

3. The Pessimistic estimate is established on a worst-case scenario. It is the longest time required to complete an activity.

Utilizing PERT, the weighted average for the duration estimate of each project activity must be computed by the formula below (Passenheim, 2009):

$$\text{PERT weighted average} = \frac{\text{optimistic time} + (4 * \text{most likely time}) + \text{pessimistic time}}{6}$$

Figure above: **PERT Formula**

The example above will clarify the difference between PERT and the Critical Path.

The critical path method shows activity duration for two days while the PERT method displays activity duration of three days. This shows that PERT can handle uncertainty in activity completion times

The major merit or benefit of the PERT is that it endeavours to address the associated risks with duration estimates. However, the disadvantage of this method is the inability to assess risks for multiple duration estimates.

6.6 Time Management

Time is basic and crucial to the individual and business life of everybody. A project manager invests a lot of time interacting and connecting with other people who are executing the project. Thus, it is essential that time is spent in a profitable and efficient way. A project manager must be careful to ensure that he focuses on the main task especially those that implement the project objective.

How time is utilized should be well analyzed that it may be maximized and lead to higher productivity.

There should also be a frequent record of how individual time is spent.

A daily log should be collected of about three weeks' activities showing how much time is spent doing every action, who was included, and what was proficient. Agenda can be categorized, for example, phone, meetings, unplanned guests or exceptional demands.

A study of time distribution by category will enlighten the project manager to figure out where his or her time is spent most, so corrective actions can be made. It is often simpler to decrease a category of high cost or expense of time by a little amount than it is to lessen a category of low use of time.

Basic time wasters of project managers are unnecessary phone chats, ineffective meetings and unproductive calls. Although the phone is fundamental for a manager to play out his or her work, it can serve as a source of distraction if not properly managed.

There are times when calls should not be answered to allow for implementation other tasks and duties. A secretary, personal assistant, or voicemail can block calls to aid phone call management.

Managing Project Meetings: Meetings are required for project management.

The best approach to conduct an effective meeting is to prepare an agenda or outline and convey same to all participants in advance. This outline or agenda enhances focus and direction and ensures that participants follow the required information in an organized manner. The table below displays some common time wasters.

COMMON TIME WASTERS

1. **Attempting too much at once**
2. **Inability to set and keep priorities**
3. **Failure to say no**
4. **Lack of goals and objectives**
5. **Inability to delegate**
6. **Special requests**
7. **Impromptu visitors**
8. **Procrastination on decisions**
9. **Ineffective meetings**
10. **Unproductive telephone calls**

The project manager must set priorities and build up a framework to spend his or her time wisely. Less fascinating duties can be scheduled at the pinnacle of one's energy and motivation.

Appropriate delegation should be done, and an analysis of work to determine what is to be combined or eliminated should be carried. Priority is to be placed on long-term items, as opposed to short-terms as people easy carry out planned activities than unplanned ones.

A few things to note

The critical path is an essential result of a project plan. It is imperative to comprehend the critical path to know where you have critical activities and where you don't. There may be a lot of activities that end up running late, yet the general project will, in any case, be completed in due time since late activities are off the critical path.

Bill remarked, so much to recall.

Rich Content indeed commented Mark.

QUESTIONS

CHAPTER 6: PROJECT PLAN WITH TIME MANAGEMENT

1. .... And .... are ways to approach project networks

2. Every activity must have a peculiar identification number
   a) True b) False c) No idea

3. Conditional statements are not to be utilized as a part of project networks
   a) True b) False c) No idea

4. The *backward pass* calculation does not start with the last project activity on the network
   a) True b) False c) No idea

5. The three-point duration estimate of an activity are ...., ..., and ....

6. Basic time wasters of project managers are unnecessary phone chats, effective meetings and unproductive calls
   a) True b) False c) No idea

7. The critical path is an essential result of a project plan
   a) True b) False c) No idea

ANSWERS

1. Any two of (Pert, Gantt Charts, Activity on Arrow, Activity on node network, Critical path method)
2. A
3. A
4. B
5. (optimistic estimate, Most Likely estimate, Pessimistic estimate)
6. B
7. A

CHAPTER 7

STRATEGIC CONTRACT SELECTION

What is a Contract?
Right Contract Selection
Contract Types
Contract Documents
Selecting the Contractor
Sub-Contracting

7.0 STRATEGIC CONTRACT SELECTION

Good morning guys, said Paul.

Good morning sir, replied both Mark and Bill simultaneously.

Today, our emphasis will be on selection of contracts.

In this chapter, we shall consider the following:

1. What is a Contract?
2. Right Contract Selection
3. Contract Types
4. Contract Documents
5. Selecting the Contractor
6. Sub-Contracting

Once a project manager is appointed at the early stage of a project, one important issue to be sorted out is the contract strategy that best suits the project objectives.

Contract strategy involves the selection of contractual and organizational policies needed for the execution of a specific project. The development of the contract strategy consists of a complete assessment of the available alternatives for the management of project design and construction to maximize the possibility of accomplishing project objectives.

The scope of such contracts is extremely wide, from an easy purchase of standard article to multi-million-pound projects. The scale and complexity of the contract matter vary consequently.

A proper contract strategy for a project involves the following key decisions:

- The setting of the project objectives and constraints
- The selection of the right project delivery methodology
- The selection of a proper contract type/form
- Contract administration practices

7.1 What is a Contract?

A contract can be defined as: "an agreement made between two or more parties which are enforceable by law to provide something in return for something else from a second party"

(Elbeltagi, 2009). Contracts could be quite simple and short or lengthy and more complicated legal documents.

As soon as a contract is properly developed, it is legally binding upon those involved. The parties involved are expected to implement the several obligations they have decided to undertake, as stated in their mutually agreed set of contract documents.

It is therefore fundamental that a contract protects the client and contractor. It is also of importance to note that not all agreements are contracts. However, a contract is an agreement enforceable by law.

Characteristics of a Contract Agreement

A contract is expected to possess the following features before it can be termed a contract (Elbeltagi, 2009).

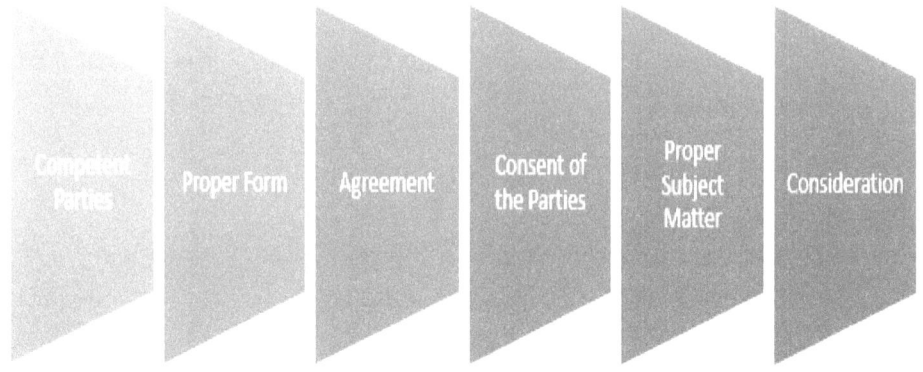

1. Competent Parties: For an agreement to be a contract, it requires at least two competent parties. And one of the parts must have a certain legal standing before it is considered competent.

2. Proper Subject Matter: A clear definition characterized by the rights and obligations of each party is the first requirement of the subject matter for a complete contract. While

the second requirement is to ensure that the contract is in line with the law and not in violation.

3. Consideration: This is a phenomenon where something is given for something. There must be a legitimate and significant thought given to both parties, and consideration must likewise be conceivable.

4. Agreement: A shared agreement must be in place to establish a valid contract. An agreement is considered to have been achieved when an offer made by one party is acknowledged by the other second party. They must ensure enforceable law bargain.

5. Proper Form: This aid in the documentation of the parties' rights and responsibilities including their contract terms.

6. Consent of the Parties: The agreement must be free from Misrepresentation, Undue Influence, Duress and so forth.

7.2 Right Contract Selection

The choice of the contract to be utilized for a construction project is made by the owner with the suggestions of the Engineer and his legal advisor. The owner's goal must be met when selecting the contract. It must also consider possible constraints the project might face ahead.

It is fundamental to adequately inform contractors and consultants of the project objectives and constraints and to note that contract selection will be affected by the nature and scope of the project where applicable.

Fig.: Steps of contracting process, adapted from (Elbeltagi, 2009)

7.3 Contract Types

There are a various types of contracts currently utilized in the construction industry. Construction contracts are categorized according to the different aspects.

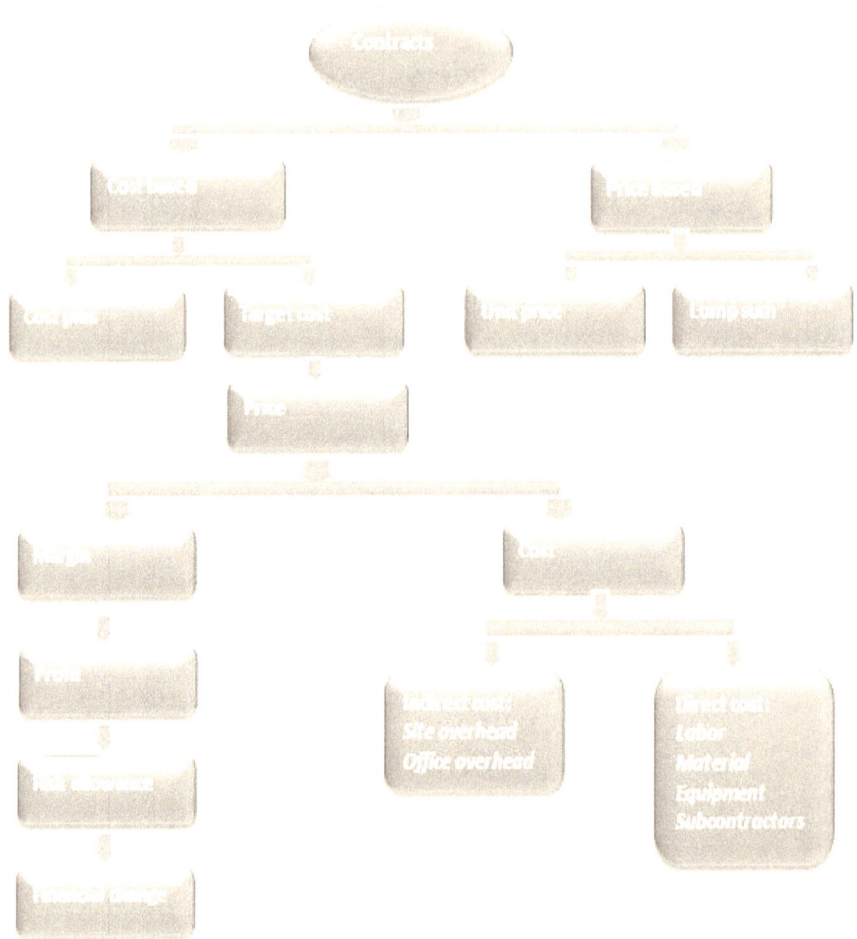

Lump-sum contract

A single tendered price is released for the completion of a task to meet the client needs by a specified date. Installment might be arranged at interims upon completion.

The contract has an exceptionally constrained adaptability for design changes. The tendered price may consist of high state of financing and possibility of high risk which may lead to trivia claims, bankruptcy or cost cutting hence, the proper risk management strategies should be implemented.

Contract final price is called a tender. One would think that a lump-sum contract would avert risks for the client, this may not be so in many cases. The lump-sum contract can expose the client to risks of not receiving competitive bids from desirable contractors who may not utilize such contract type.

This contract might be utilized for a turnkey construction. It is suitable when work is characterized in detail, constrained varieties are normal, measurable low-risk level, and the client does not wish to be associated with the management of his project.

Admeasurement contract

Admeasurement contract ensures that Bills of Quantities or Schedule of Rates covers the work items. The contractor at that point determines rates such as contingency risk against every item.

It is a pay per use thing as Installment is paid monthly for all work finished amidst the month. The contract offers a platform for the client to incorporate possible changes in the work defined in the tender documents.

The contractor can claim extra payment for any adjustments in the work content of the contract. Claim resolution is extremely difficult since the client is uninformed of actual cost or concealed contingency. Variation and claims often increase the tender price.

There are two types of admeasurement contract utilized:

- Bill of Quantities
- Schedule of rates

Bill of Quantities Contract: In this kind of contract, tenders are against the evaluated amounts of work of each item. The quantities are re-measured over the span of the contract, and the contract price is adjusted in line with the tendered rates value.

Schedule of Rates Contract: In this case, the contract cost is derived by measuring the man-hours, plant-hours and the material quantities consumed, and afterwards pricing is done at the tendered price. This type of contract uses estimation to determine amounts of work, conceivably with upper and lower probable limits, hence not uncommon to quote a variety of prices for labour, plant, and materials.

This contract is best appropriate for repetitive works. The admeasurement contract is easily understood and utilized. It can be applied when design and construction need overlapping, where there are minimal or no changes and low-level risk environment.

Cost-reimbursable contract (cost-in addition to contract)

The contractor is compensated for actual cost in addition to a special charge for head office overheads and gain, and no payment for risk. The final price depends on the degree of changes to which risks appear since design changes involve a high flexibility level.

The contractor must avail his accounts and records for examination by the client or by some agreed third party. This contract has no direct monetary incentives for the contractor to perform proficiently. It might be utilized when it is suitable for design to continue simultaneously with construction and when the client wishes to be engaged with contract management.

Other contract types include

- Target cost contract
- Time and material (T&M) contracts

7.4 Contract Documents

A contract is also characterized by the contract documents, which are developed from the tender documents. Logically, these documents record consists of the following headings:

1. Input from the client (Description of task).
2. The output of the contract (details, results to be accomplished).
3. Price for the contractor's contribution.
4. Procedures and Responsibilities (resources provided, liability, payment conditions, schedule, change procedures, and so forth).

Contract documents are normally outlined in the following sequence

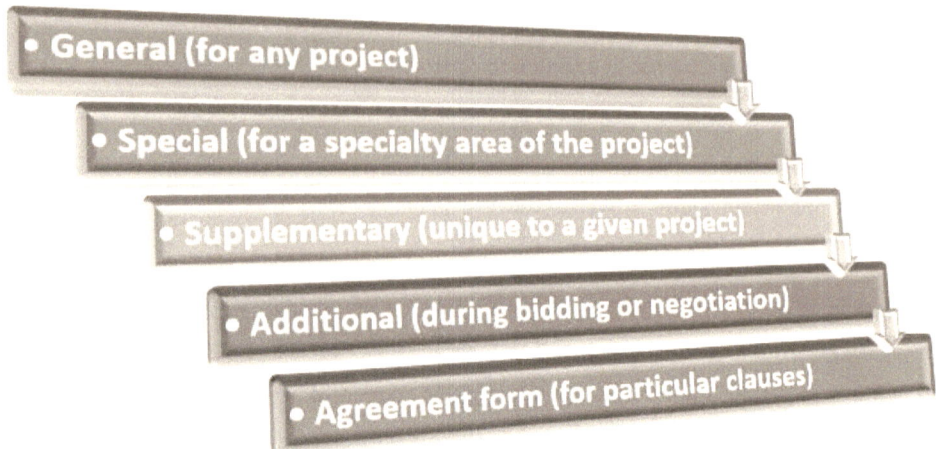

- General (for any project)
- Special (for a specialty area of the project)
- Supplementary (unique to a given project)
- Additional (during bidding or negotiation)
- Agreement form (for particular clauses)

The complete contract, for the most part, comprises of these documents:

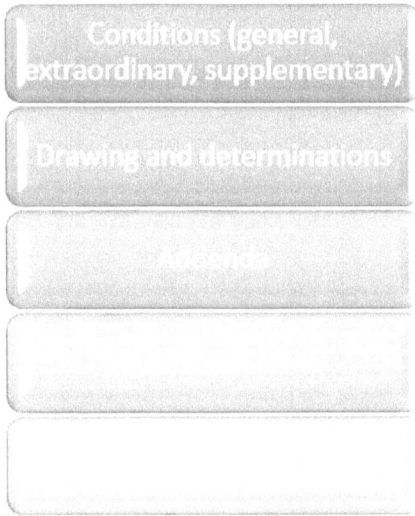

The most important of these documents from the legal perspective is the **agreement**. It is often called the contract. Since such a variety of document is incorporated as contract documents, the agreement is the better term for this specific one.

The agreement is often divided into three sections of information. The initial section is a short introductory paragraph consisting of the agreement date, parties involved and the task each party agree to undertake.

The second segment contains the components of a contract and outlines the work to be accomplished, and the last section affirms the agreement and provides space for appropriate signatories. Thus, the information below displays the components of the agreement.

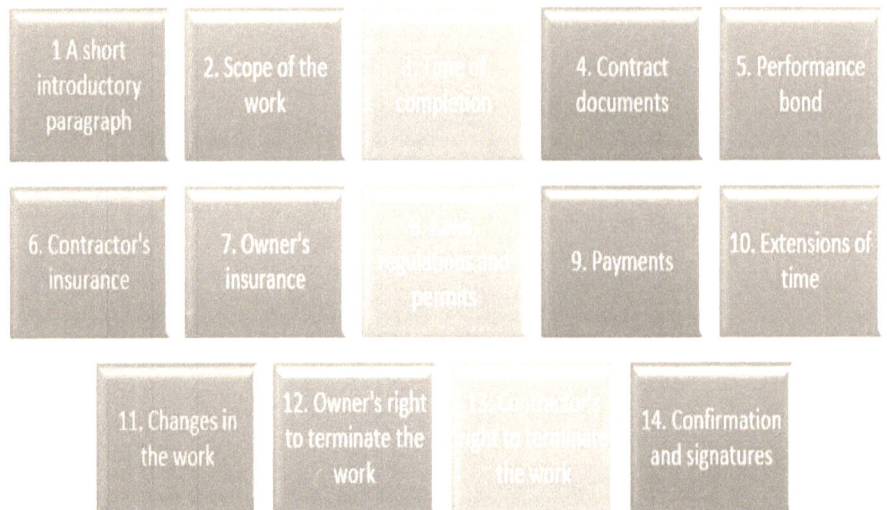

Figure: Components of agreement, adapted from (Elbeltagi, 2009)

7.5 Construction claims

A construction claim is a demand for payment or time extension to which the contractor considers him as/herself entitled. Claims are required under three types

1. Extension of time only
2. Additional cost
3. Both extensions of time and extra cost.

The principal reasons for construction claims may include:

- Different ground and/or site conditions
-
- Delays in approval and examining work
- Unforeseen events and disasters
- Work postponement by the client

7.6 Selecting the Contractor

The success or failure of a project can be affected by the right or wrong choice of selection of key personnel and organizations involved in a project. Most often than not contractors have to acquire jobs by the competitive bidding process.

This is a required procedure by law for public projects, which has been the biggest share of all projects, aside those that show up during war or cataclysmic events. Under this procedure, a basic quantitative measure is utilized to grant the bid to the "lowest responsible bidder", hence potentially obtaining the most reduced construction cost.

The procedure, be that as it may, has its downsides, including:
(1) Ignoring vital criteria, for example, contractor's strength and experience;
(2) Conceivably causing construction delays and issues if the contractor offers cost beneath the bid cost to win the job
(3) Add to adverse relationships between the owner and the contractor.

The competitive bidding process involves three primary stages:

The owner ought to have the finished design and a bid package arranged with all relevant design information for a project to be announced. At that point, the owner announces a general call for bidders or sends a limited invite to a list of pre-qualified contractors.

The owner can lessen potential construction issues by maintaining a strategic distance from unfamiliar contractors who purposefully decrease their bids to win jobs, especially if the project requires a specific experience. It is therefore expedient that owners have a list of qualified contractors with who possess effective experience.

7.7 Sub-Contracting

Most construction projects sub-contracted some work to *speciality* contractors, known as subcontractors. The greatest piece of the work is subcontracted on building projects, with a lesser amount subcontracted on heavy construction projects.

Construction contracts, for the most part, have clauses relating to subcontracting. Such clauses regularly restrict the measure of work to be subcontracted and by and large give the client the privilege to endorse sub-contractors.

The contractor who utilizes subcontractors to carry out part of the works must be responsible for their workmanship, performance, and general conduct on the contract. Any correspondence on these aspects should be carried out between the primary contractor and the client. The contractor is to ensure this particularly.

Any limitations or restrictions on subcontracting should be well stated in the tender records. If the client wishes a specific subcontractor to carry out part of the work, he may make such nomination.

When work of a specialized nature is required, such nomination of subcontractors should be welcome. It is often a common deed for customers to place other limitations on subcontracting simply by affirming the main contractor to a list of the endorsed contractor, though this happens at the tender stage.

Bill said 'Paul, so much to know on contract selection'

Yes, so much information, commented Mark.

I understand guys, but you must remember that having the appropriate legal documents will aid the parties involved in the project to execute their obligations with understanding and less strife; replied Paul.

QUESTIONS
CHAPTER 7: STRATEGIC CONTRACT SELECTION

1. Once a contract has been developed, it is legally binding
   a) True b) False c) No idea

2. Proper form and disagreement are characteristics of a contract agreement
   a) True b) False c) No idea

3. Tender meetings are part of a contracting process
   a) True b) False c) No idea

4. A Contract final price is called a tender
   a. True b) False c) No idea

5. Contract documents are normally outlined in the following sequence

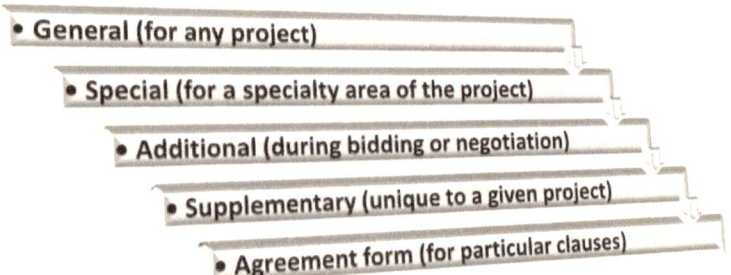

   a) True b) False c) No idea

6. .... is the most important document of a contract from the legal perspective

7. Extension of time is a component of agreement
   a. True b) False c) No idea

8. It is not compulsory for contractors to acquire jobs by a bidding process
   a. True b) False c) No idea

ANSWERS
1. A   2. B   3. A   4. A   5. B   6. Agreement   7. A   8. B

CHAPTER 8

Quality Concepts

Project Quality Management

 Project Quality Management Processes

Techniques for Quality Control

Business process re-engineering (BPR) and Total Quality Management (TQM)

8.0 PROJECT MANAGEMENT AND QUALITY

Paul entered the conference room with his bag in his hand and said, good day guys, what a chilled morning. Please, have your sit and get yourselves a cup of coffee while we dive into a new topic today.

Thank you sir replied both Bill and Paul.

Paul continued, our topic for today is Project Management and Quality and our focus areas shall include:

1. Quality Concepts
2. Project Quality Management Processes
3. Techniques for Quality Control
4. Business Process Re-engineering (BPR) and Total Quality Management (TQM)

8.1 Quality Concepts

In the 1950s, quality was perceived to be the art of examining products that had just been released with the objective to isolate the good items from the defective or blemished ones.

However, present business methodology anticipates imperfections or defects and plan to prevent this rather than examining them; i.e., it is difficult to produce a good product by inspecting after production.

There is a need to concentrate on procedures to guarantee things are done well the first time, unfailingly, and in a cultural environment where everyone is quality centred.

Nevertheless, in the quest for competitiveness in the marketplace, project teams regularly find alternatives to cut expenses and accelerate schedules, even though this ordinarily increases errors, the need to modify more creates greater workload for the project group, and a "quality meltdown."

What is Quality?

Quality infers meeting specifications or requirements. However, it additionally implies more than that. While meeting project specifications will keep a client from prosecuting a contractor, specifications alone cannot guarantee that the client will be satisfied with the final product or the contractual worker will be appreciated or win repeat business (John M. Nicholas and Herman Steyn, 2012).

Preferably, a project should look beyond specifications and tries to fulfill client desires— including those not articulated; it aims is to exceed customer's expectation more of "extreme customer service" approach. A typical shortcoming of project managers is the assumption that the client needs, desires, and expectations will require little effort to implement.

Fitness for Purpose

The expression "quality" infers that an item or deliverable is fit for the intended purpose. Fitness regularly includes an extensive variety of criteria, for example, execution, safety, maintainability, logistical support, and no destructive ecological effects.

For example, a costly bit of attire or product may have unrivalled style, material, or workmanship, the client will likewise consider its monetary value and whether it is adequately priced for the intended purpose.

The project manager must ensure that the multiple phases of producing a product are balanced to reflect complete quality and not just improve a single phase of producing an item. S(he) must adequately combine fitness for purpose, customer fulfilment, monetary value or key benefit to the organization or clients

Absence of Defects

Another meaning of Quality is the absence of Defects, which is the reason individuals habitually relate quality with a defect. A defect is a phenomenon in which an item or product does not conform to the intended desire of a client.

One approach to accomplish quality is to recognize quality at the earliest opportunity which involve distinguishing and correcting a defect as early as feasible under any circumstance.

Generally, the more the delay to detect nonconformity, the more expensive it is to rectify or remove. It may be less difficult and cheap to fix a defect in a segment part, but costlier to fix it after the part has been assembled, and costing more after the assembly is part of a bigger framework or system.

The defect or deformity is most costly when it results in product failure or system malfunction while in use by the client. Nonetheless, "absence of defects" requires qualification, and the assumption that zero deformities are equal to high-level quality is not always true.

A quality project is one that satisfies different requirements and necessities. Investing excessively to a specific requirement, for example, eradicating all defects may affect the fulfillment of other prerequisites that are more important.

For instance, the requirements for projects include time, cost, and execution. In a situation where the schedule must be sustained, attempting to eliminate all deformities can be costly. The client may prefer to minimize expenses and hold to the schedule instead of removing all deformities.

Obviously, some cases will warrant complete removal of every defect. Even the most minor defect in an air traffic control system or artificial human heart can bring about damage or death toll.

The fact of the matter is, it relies on the client. Often, the client would prefer products or services finished on time, with a lower cost, and minor deformities to one completed late, at a higher cost, however without any defect.

Adequate Quality

In eliminating defects, the focus is on those deformities that would keep the system from meeting its most critical necessities. This is the idea of "adequate quality."

This is the criteria when priorities on execution prerequisites like time, and cost meets all the requirements and constrain the project team to narrow down to essential tasks.

The art of providing the best possible quality takes a high investment cost, although most clients may not observe the disparity between the best possible quality and pretty good quality.

The client, obviously, must have the capacity to judge what is "sufficiently good," and to do that, there must be a continuous update about the progress of project, challenges, expenses, and schedule.

In the perfect case, everybody on the project group adds to quality;

Everyone:

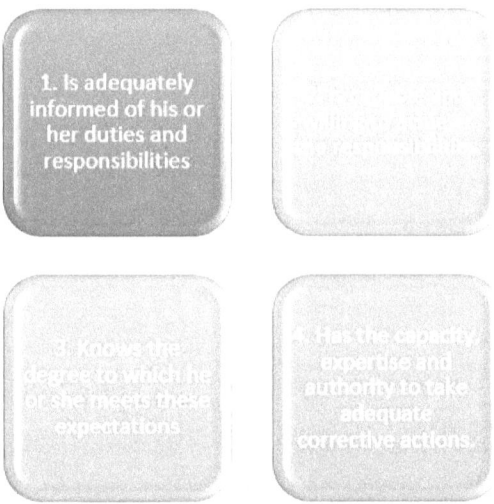

Such conditions require quality-centered leadership, training, and motivation factors. Once everybody begins contributing in their various capacity, quality enters an automatic procedure thus requiring less influence from the project manager.

Quality Movements and Progress

Dr W. Edwards Deming, a significant American expert, proposed a "quality revolution" term which began in Japan in the 1950s. It consists of continuous improvement, skills development, levels of leadership, eradicating reliance on inspections, dependence on single-source (as opposed to many-source) suppliers, and utilization of sample testing and statistical approach.

From that time forward, various other quality developments have passed through our ears. Some could be portrayed as prevailing fashions. The most enduring and well-known development since the 1980s is total quality management (TQM). TQM is a set of system strategies.

It is more of a mentality, a voracious approach to enhancing the total effectiveness and competitiveness of an organization. The key components of TQM recognize the mission, objectives, and goals of the organization. It acts consistently in line with these objectives and goals, and also concentrate on consumer satisfaction.

TQM covers the total organization including, teams of forefront personnel and evident support from top management. Quality issues are methodically identified and resolved to enhance processes constantly. Project reviews and close out sessions serve this purpose in the management of projects.

TQM is also complemented by another management philosophy termed 'just in time (JIT) or lean manufacturing/production. Lean manufacturing can detect that quality issues begin from "broken procedures." It gives strategies and methods to analyze processes uncovering and removing non-value added waste in processes (John M. Nicholas and Herman Steyn, 2012).

The lean methodology incorporates simple techniques to enhance quality, reduce expenses and lead times.

The challenging aspect of actualizing lean is that it requires building up a culture where staffs in every place have the skills and authority to constantly improve their processes—an uncommon model for many organizations.

The concepts of lean manufacturing began at Toyota in the Toyota Production System (TPS), and have been effectively received across the nations in the world. In auto and electrical industries, lean production is now embraced.

In engineering projects, lean manufacturing methodologies are implemented for the development and construction of products.

Another influential quality development is Six Sigma which commenced in the 1980s at Motorola and later propagated by General Electric. Supporters of Six Sigma postulates that it produces a more organized way to approach quality than TQM.

The expression "Six Sigma" refers to a scenario where the normal distribution percent of the population falls within 99.99966% −6 δ to +6 δ of the mean, where " δ " is the standard deviation. If the quality of a process is controlled to the Six Sigma standard, there will be maximum 3.4 parts per million defects in the process—close to flawlessness or say perfection!

However, the Six Sigma approach is more than measurements, but a concept for diminishing process variability. It incorporates a five-stage process for enhancing existing procedures, and another five-stage process for designing new procedures and products.

These processes are for Six Sigma quality levels. The first procedure is called

DMAIC (Define, Measure, Analyze, Improve, Control), and includes the means of defining the best measures to improve a procedure, executing those measures, following up on the measures, and lessening defects to meet specifications.

The second procedure, which concentrates on configuration, utilizes a comparable procedure called

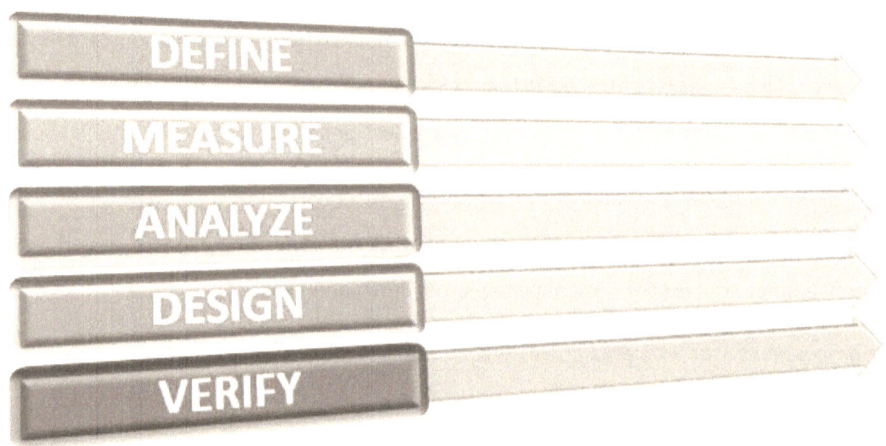

DMADV—Define, Measure, Analyze, Design, and Verify.

In projects, the Six Sigma approach ensures the definition of clear expectations that identify with the mission of the organization and are approved by management.

The DMAIC procedure is often exploited as the project methodology whereas, the five stages of the process define the different phases of the project.

8.2 Project Quality Management

Quality of a product begins with a plainly defined product or system prerequisites of the contractor and client as agreed. In situations where the client gives pre requisites or specifications that appear to be unfeasible, the contractual personnel should review with the client and modify to accomplish the desired objective.

The final specification agreement ought to reflect the client's desires, product's fitness for the intended purpose, and any other compromise negotiated. Nevertheless, comprehensive specifications ought to be incorporated into the project scope definition.

Project quality management incorporates quality management procedures and strategies to lessen the risk of products not meeting prerequisites.

8.3 Project Quality Management Processes

Project quality management comprises of three procedures:

Quality planning guides future quality agenda; it sets the prerequisites and benchmarks to be met. It also covers important processes to meet them.

Quality assurance implements the planned quality activities and guarantees project activities to meet quality specifications and end-item requirements.

Quality control ensures that quality assurance agenda is carried out in line with quality plans, and the specifications and requirement are met. You can consider quality control as the "medication" to dispose of existing deformities while quality planning and assurance as termed "the healthy lifestyle" to avoid deformities initially.

As depicted in the Figure above, project quality control overarches quality planning and quality assurance. The part of quality control is to guarantee that quality confirmation occurs as indicated by the quality arrangement. Quality affirmation intends to guarantee there are fitting quality guidelines for a project and to exploit taking in circumstances from finished projects for a consistent change of future projects.

Quality Planning

Quality Planning ought to give confidence that all things required to guarantee quality have been thoroughly considered. It has two angles:

(1) Development and establishment of project quality management techniques and policies for the whole organization

(2) Development and establishment of a quality master plan for each project.

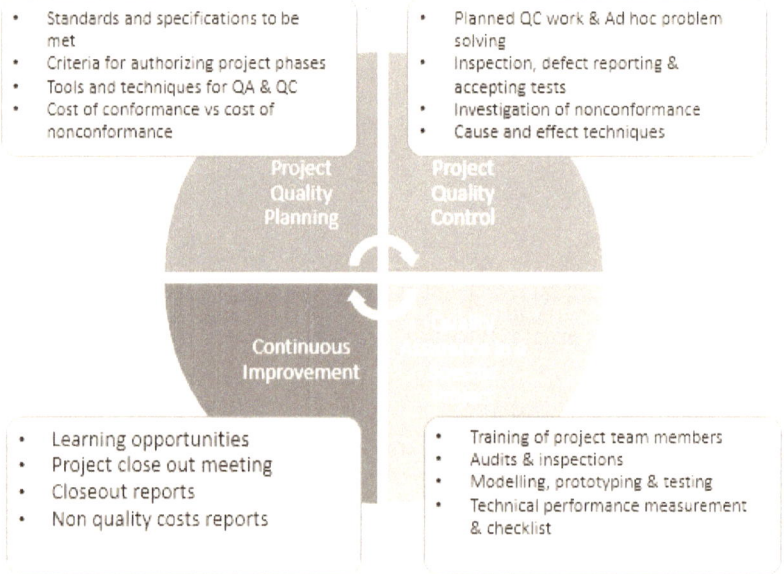

Figure: The project quality management process adapted from (John M. Nicholas and Herman Steyn, 2012).

The task for setting up organization-wide policies and strategies to enhance project quality management ordinarily falls on functional managers, particularly the quality manager. Projects regularly utilize current quality standards in existence in the organization, for example, the ISO 9001 standard, in a quality management system.

In circumstances involving design and development of projects, this standard recommends that an organization shall set adequate procedures for

(a) The design and development stages;

(b) The fundamental reviews, verifications, and approvals fitting to each of the stages

(c) The duties, tasks and authorities for the stages.

Project planning process always incorporates quality planning. The prerequisites to be met to approve the next project stage should be determined in the quality management plan. It should also indicate the quality approach to be utilized.

The quality management plan should also indicate how the project group will execute the organization's quality policies. This should come together in the plans for human resources, procurement, communications, risk, Health and safety, and so on in the project master plan.

Quality Cost

Since quality is constantly identified with the monetary value expanded, quality planning is expected to consider the expenses and advantages of quality activities. A cost–benefit analysis is completed to evaluate and legitimize proposed quality activities and to compare quality expenses and control actions with the savings or benefits from less or eliminated deformities due to those activities.

Money spent on quality assurance and control ought to be justified regarding risk reduction of not meeting specifications.

Quality cost can be classified as

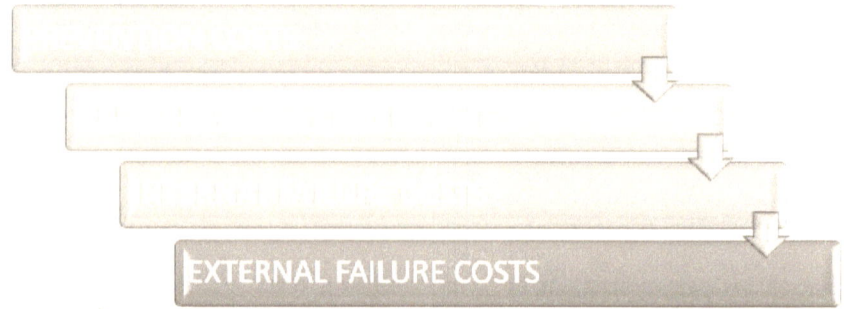

1. Prevention: training expenses, design reviews, and any action directed at preventing errors; including the cost of quality planning.

2. Appraisal and control: costs of evaluating products and procedures, including product audits, reviews, tests, and inspections.

3. Internal failure: costs related with deformities discovered by the manufacturer; costs for scrap, rework, and retest are likewise included

4. External failure: costs acquired due to product failures after delivery to client; including costs for replacements, guarantee repairs, liability, lost sales, and damaged reputation.

While the quality cost of an organization with a sound quality management system can be as little as 2 percent of the organization's revenue, they can surpass 20 percent for an organization with a low-quality management system.

It is therefore wise to invest resources into a decent quality management system such as spending more on audits, design reviews, modelling and testing to spend less on internal and external failures.

External failure costs occur after the project is finished in many projects, while the prevention costs, appraisal, and control are inquired amid the project.

Therefore, costs for prevention, evaluation/appraisal, and control ought to be estimated, incorporated into the project plan, and covered in the budget for the project. These include the many costs the project manager must justify to management and the client.

Quality Assurance

Project quality assurance identifies with the implementation of the project quality management plan. It decreases the dangers associated with not meeting necessary performances.

Quality assurance includes the following:

1. Actions performed in a specific project to guarantee that necessities are being met and that the project is being implemented in line with the quality plan.

2. Exercises that add to the consistent change of present and future projects, and to the project administration development of the association.

Quality assurance ensures that all things needed to guarantee the proper quality of project deliverables are done. (John M. Nicholas and Herman Steyn, 2012).

Continuous Improvement and Project Post-Completion Reviews

It would not be unusual to state that the foundation to progress is continuous improvement. Organizations involved in projects endeavour to constantly enhance their specialized operations and managerial procedures by conducting a formal close out or post-completion review for each project.

The post-completion review reveals what transpired, and to learn lessons that can be applied future projects to avoid similar errors. The reviews enable organizations to improve its technical methodologies and project management.

The project manager's duty while carrying out reviews is to facilitate genuine and constructive dialogue about what transpired—what worked and what did not—and to ensure that every individual involved is given a listening ear.

The dialogue is formally documented, and lessons learned list is generated. This procedure is fundamental for consistent progress. However, it is ignored due to loss of lose enthusiasm by individuals as the project winds down, or they become occupied by new and coming projects.

Thus, leading to repetitive errors by organizations as they are unable to gain from their experiences

Quality Control

Quality Control is the progressive process of monitoring and evaluating activities, and taking remedial action to accomplish the planned quality outcomes. The procedure additionally verifies that quality assurance exercises are achieved in line with the quality plan, and that project requirement and specifications are implemented.

In situations where deformities are revealed, the causes are resolved and wiped out. Just as the quality plan is to be incorporated with different parts of project design, quality control ought to be coordinated with alternate parts of project control.

It is the task of the project manager to incorporate scope control, cost control, advanced control, and possible risk control in maximizing quality control across the project.

Quality control can be juxtaposed to scope verification: while scope verification deals with the client receiving the project deliverables, quality control talks about adherence to specifications set by the contractor.

The quality control also ensures that deliverables are consistently meeting specifications and acceptance tests before the client receives deliverables. In situations where a product does not meet the required standard, the contractor may request a deviation or waiver that would discharge the deformed product from the specification.

A waiver relates to an unpredicted condition that is found after the product has been manufactured. It approves a temporary deformity, for example, a scratch found on the paint of a hardware product.

In a case of a deviation, a temporary withdrawal from specification is done however, it is obtained ahead of time. For instance, if a specified material is inaccessible, the contractor can apply for a deviation to permit the use of an optional material.

A deformity can likewise be classified a modification, which is a change to the standard that is considered permanent. Control activities incorporate both planned quality control activities and ad hoc problem-solving.

Planned activities could involve site inspections on a construction project, product component tests, auditing materials from a supplier. Ad hoc problem-solving deals with handling problems and risks as they emerge.

Quality of Procured Items

The quality prerequisites for most products obtained from providers are set by industry standards which are often determined by picking a supplier's price. To purchase a group of standard things such as nuts and bolts, the procurement personnel obtains quotes from various providers and picks the minimum cost.

At the point when the batch arrives, the required personnel inspects the nuts and bolts to determine their acceptability. In situations where a new item is to be developed, there is likely no industry standard yet. Therefore, the buyer will work with the supplier of such product and assist in planning to ensure quality assurance and control to meet the desired specification.

Obviously, selection of items should not be solely based on items with minimum cost but those that will simultaneously meet the contractor's requirements that both contractor and supplier work together as partners and share responsibility for each other's success.

Setting up this sort of relationship is not simple, particularly when the provider is considerably bigger than the contractor, or does not esteem the relationship or consider the contracted work a high priority.

Contractors invest highly to ensure they have continuous access to subsystems and items with appropriate quality. A contractor usually has a division within its procurement team to oversee quality assurance of all the procurement of his products as well as their manufacture/construction or development.

The aim of this division or team is to guarantee quality and perform necessary inspections and acceptance tests of products purchased to select suppliers and monitor supplier's processes adequately.

8.4 Techniques for Quality Control

Quality control includes the utilization of techniques defined in the quality management plan and other fundamental corrective activities to guarantee quality.

Inspection and Acceptance Testing of the Final Product

The testing of prototypes and models reveals design and development while, acceptance testing of the final product and deliverables verifies that the product meets the required specifications. Product termed as critical is always inspected, but those grouped as minor/incident are not.

Take for instance the automobile manufacturing, the steering and braking system of each vehicle is tested. In the case of mass production items, a few might undergo destructive tests (i.e., tested until the point when they break); while unique or in a small batch cluster are subjected to nondestructive examination and testing.

Even though testing the final products from a production process does not come under project quality management so to speak, items produced on a massive scale would incorporate the testing procedure and other quality assurance procedures to be utilized in producing the item.

The quality check should begin after production commences and to be specified or carried out by expatriates, those familiar with the characteristics of the product. Sampling inspection is carried out on items produced in high volume. This lessens the inspection cost.

Upon the outcome of these sample tests, statistical inference is made about the quality of the whole production batch or process. Mandatory sampling is done when the testing destroys the product.

Quality Control Tools

Kaoru Ishikawa of Tokyo University in the 1980s, defined the fundamental quality control tools. The purpose of these tools is to detect the causes of deformities and defects in products and processes. Their application involves recognizing sources of problems including those related to risk and resolving problems of all types. Though these tools were discovered, developed and utilized in a production environ (Kawasaki Steel Works) they are well applicable to projects nonetheless.

- Control Chart
- Cause & Effect Diagram
- Casual Loop Diagram
- Current Reality Tree
- Run Chart

THE TOOLS

Run Chart

A run chart is a graphical illustration of results observed plotted against time to uncover likely patterns or irregularities. The plot of price versus days in Figure below is a type of run graph (just a sample) that tracks variance in similar project item cost in different days.

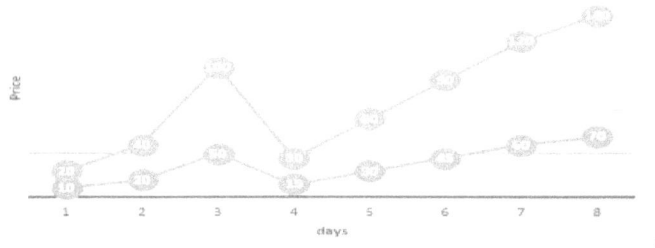

Figure: showing a run graph sample

Control Chart

Control graphs are broadly utilized for identifying process changes, controlling and tracking repetitive. For projects that incorporate the improvement of processes in production, one deliverable would be determining the relevant charts for controlling the process quality.

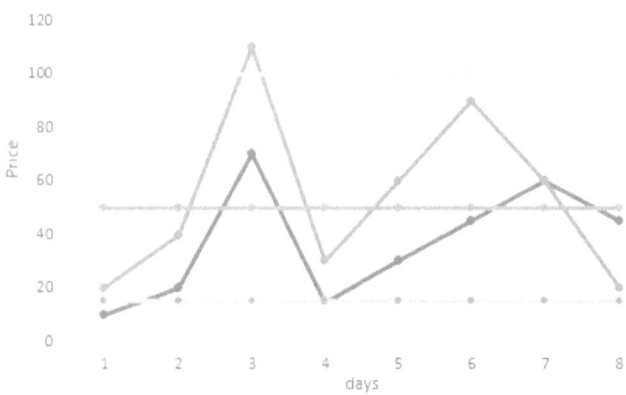

Figure: showing a control chart sample

Pareto Diagram

An Italian market analyst by the name Vilfredo Pareto, in the 19th century, defined "Pareto's Law" of income distribution as the dissemination of income and riches of a nation in a regular pattern: 80 percent of the riches is possessed by 20 percent of the population.

This is parallel to the popular "80/20 rule," and its relevance to quality cannot be overemphasized. Quality expert Dr. Joseph Juran in the late 1940s postulated that most of the deformed products result from a relative few cause, thus reasonable to identify the few relevant causes of the bulk of deformed products and to ensure removal of same, especially for economic purposes.

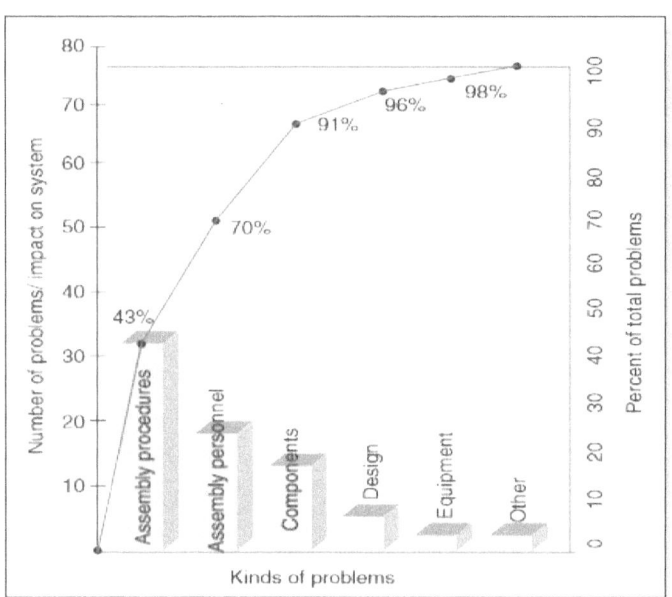

Figure: Pareto diagram adapted from (John M. Nicholas and Herman Steyn, 2012), page 340

The figure above depicts a Pareto diagram sample. At the diagram base is a histogram displaying the kinds of problem against the problem sources; the diagonal line crossing the figure is the combined or cumulative impact of the problems (corresponding to the right-hand scale).

As displayed, the initial kind of problem represents 43 percent of the problems; the combination of the initial and second problems represents 70 percent. This simply implies that resolving only the first two problems eradicates 70 percent of the problems.

In carrying out projects, sources of most problems can be easily identified using the Pareto analysis, especially sources in need of most attention.

Cause and Effect Techniques

Cause and Effect Diagram

Quality risks and problems are often best resolved through the combined knowledge and experience of project team members. Group members meet to conceptualize thoughts and brainstorm regarding problems, and these thoughts result in ideas which are recorded on a cause-and-effect (CE) diagram (also known as fishbone or Ishikawa diagram).

It is a schematic diagram for outlining the causes for a specific effect in a logical way. The figure below demonstrates a CE chart to decide why a problem occurs.

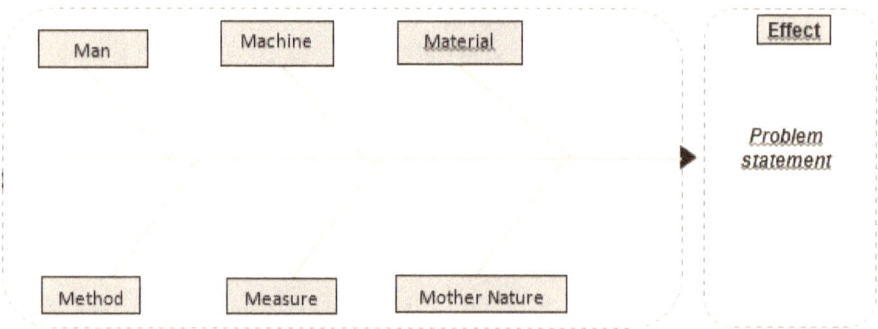

Figure: Fishbone diagram

As the project team members think through the problem and produce relevant ideas, each suggested cause is assigned to a specific branch as shown above.

Causal Loop Diagram

A causal loop diagram can be utilized to depict the causes for a certain problem or the influence of a few factors in a diverse system on one another. In the figure below, variables in the chart are linked by arrows to indicate cause–effect relationships.

The positive and negative signs reveal the direction of change of the variable as the arrow head or tail changes. A positive sign displays a reinforcing impact—the variable at the arrow head increases when a variable at the arrow's tail increments.

A negative sign demonstrates a negative impact—the variable at the arrow's head reduces when the variable at the arrow's tail increases.

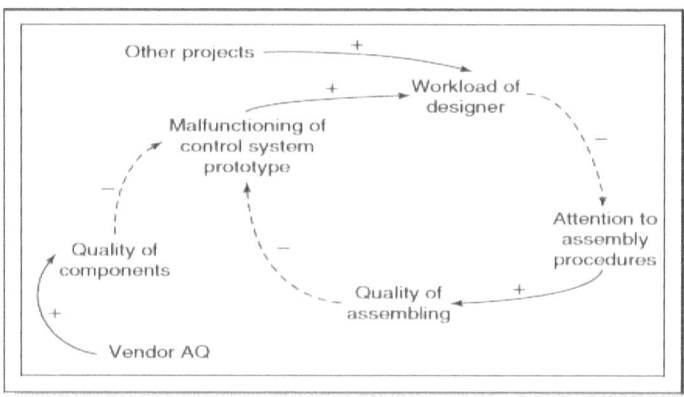

Figure: control loop diagram adapted from (John M. Nicholas and Herman Steyn, 2012), page 340

For instance, in Figure above, as the project rises, the designer's workload soars, which could lead to a reduction in attention to detail required by the designer.

Current Reality Tree

Another alternative to analyzing cause and effects is current reality tree (CRT). The technique begins with an identified /observed undesirable effect (UDE) or symptom; this outcome is utilized to identify similar connections between the UDE and other undesirable effects.

A CRT considers whether the causes identified are sufficient to bring about the specified UDE unlike that of causal loop diagrams or CE. This logical strategy entails that assumptions surrounding situation or circumstance be identified, and to reveal as many dimensions of the problems involved which ultimately identify the fundamental causes or problems.

For instance, assuming a possible source of a faulty control system is the system assembly procedure. The figure below demonstrates the CRT for the situation; it is understood from the base to the top as follows: entities 100, 200, 300, and 400 are causes for the UDE numbered 500. The oval around the arrows shows that these four entities are sufficient to have resulted in UDE 500.

Similarly, entities 100, 500, 600, 700, and 800 are sufficient to have caused UDE 900. The CRT technique requires more effort than basic CE analysis, thus preferable to be implemented for issues or problems considered more severe (the 20 percent of the causes that lead 80 percent of the problems).

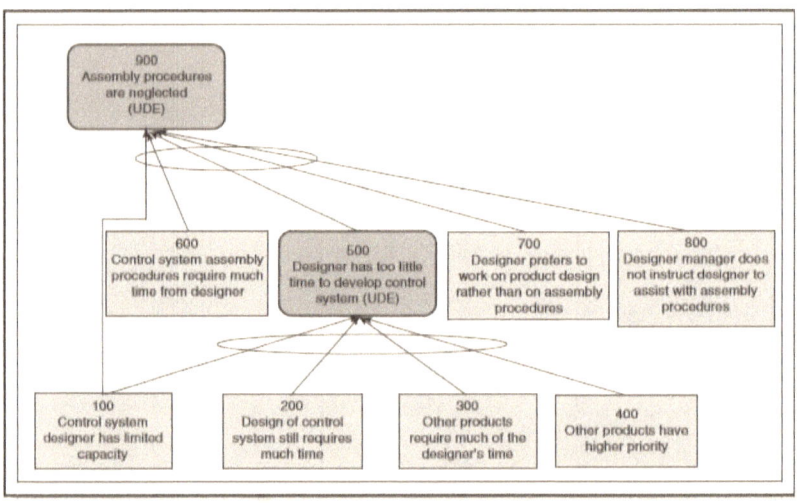

Figure: CRT sample adapted from (John M. Nicholas and Herman Steyn, 2012), page 342

8.5 Business process re-engineering (BPR) and Total Quality Management (TQM)

Total Quality Management can be described as a continuous improvement process strategy. It was popularly embraced in the 1980s, and by 1990s many organizations began to introduce Business Process Re-engineering (BPR) by testing with more radical change approaches (Smith, 2010).

BPR and TQM share many features, such as process orientation, customer focus and commitment to improved performance.

Nevertheless, these two processes have several differences as shown below

Table showing the fundamental difference between BPR and TQM, adapted from (Smith, 2010).

S/N	Similar Characteristics	BPR	TQM
1	Role of technology	IT enabler	Statistical Control
2	Nature of Implementation	Top-down	Bottom-up
3	Main Control Measure	Cost of Process	Cost of Quality
4	Process Orientation	Cross-functional	Functional
5	Type of Change	Revolutionary: a new way of doing business	Evolutionary: a better way of doing a job
6	Method	Redesigns the business process	Adds value to existing process
7	Improvement	Dramatically	Incrementally (Continuous)
8	Risk	High	Moderate
9	Scope of Change	Core business	Organization processes
10	Time of Investment	Time for training and culture change	Time for establishing information systems and organizational structure

The implementation of TQM and BPR is of current challenge among many as to determine which comes first? Many project organizations re-engineer processes before introducing a quality system to ensure that the latest IT enablers are incorporated in the system.

However, if a quality system is developed without inadequate investigation of the required process, there is the likelihood that the system will require major development as the project progresses.

Bill took a deep breath and said, I perceive that quality management principles and tools can be utilized across various project.

Yes please, it is critical to the success of any project. Paul responded.

QUESTIONS
CHAPTER 8: PROJECT MANAGEMENT AND QUALITY

1. Zero deformities is equal to high-level quality.
    a) True b) False c) No idea

2. Develop is part of the Six Sigma quality levels
    a) True b) False c) No idea

3. ………. And ………… are project Quality Management Processes

4. Appraisal and Control costs are part of quality cost
    a) True b) False c) No idea

5. ………… is a graphical illustration of results observed plotted against time to uncover likely patterns or irregularities
    a. Run chart b) control chart c) current reality tree

ANSWERS
1. B
2. B
3. Any of these (Quality planning, quality assurance, quality control)
4. A
5. A

CHAPTER 9

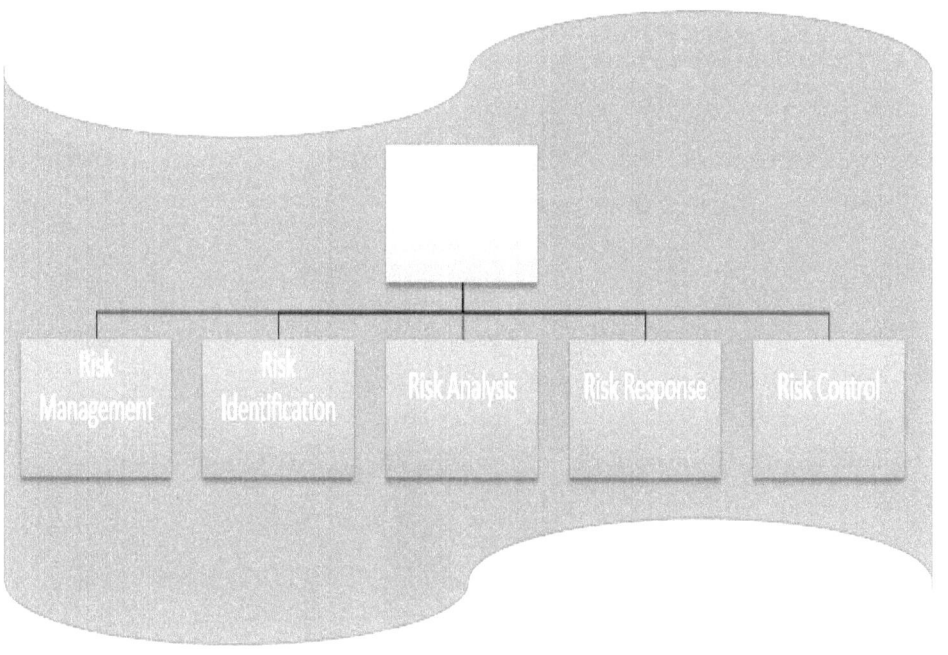

9.0 RISK MANAGEMENT

9.1 Introduction

Hi guys, today we shall discuss a very interesting topic we face in our daily lives, said Paul as he took out his Ipad from his bag.

Mark and Bill wondered what this could be and just before they could speak, he broke the silence and said 'it is a four letter word guys, our topic is on RISK today'

Wow, responded Mark. Go for it Bill, am pumped to hear more.

Risk management plays a major role in project management, said Paul. Ineffective risk management plan could have an effect on the project.

Definition of a Risk

Since uncertainties are part of risks, it is relevant we know what this means before getting into risk proper.

Threats are events that negatively affect any outcome.

- Opportunities are events that positively affect results;

- Uncertainty includes the total scope of positive and negative effects.

We are often involved in various risks in our daily life, for instance, we could be harmed in a vehicle accident if the safety belts are not secured. In the event that we do smoke excessive number of cigarettes the likelihood of dying of cancer is significantly higher than a non-smoker. It is not our temperament for example, to consider risk management involving situation which may influence us however, risks unquestionably shape our practices. In situations where we need to cross the road, guardians have instructed us to look both ways before setting a foot on the road.

Additionally, in each project, there is the likelihood for dangers and benefits which may influence the completion of a project. Many are of the opinion that a risk is just a problem waiting to happen, however, this is not completely true. A risk is not a problem until it truly manifests. It is progressively an acknowledgement that a conceivable problem may happen in the future.

The occurrence of risks can be avoided when the project manager initiates appropriate countermeasures. Project risks are those which can make postpone a project or impact same to surpass the initial budget. The field risk management manages both – positive and negative – parts of risk. In most cases, the focus is usually on the negative consequences of risk management as the project team is only concerned with the safety aspects of the project.

9.2 Risk Management

Risk management is a system to limit the adverse effect of a financial loss by:

1. Identify potential sources of loss

2. Measure the financial impact of an occurring loss

3. Limiting actual financial loses or consequences with the implementation of controls

The aim for observing all project risks is to maximize the value of each single activity within the project. The potential advantages and dangers of all factors associated with these exercises must be ordered and well documented. Adequate awareness of the relevance of the risk management process increases the probability of success and lessen that of failure (Passenheim, 2009)

Identification of risk is not exclusively the responsibility of the project manager. It should include all relevant stakeholders including

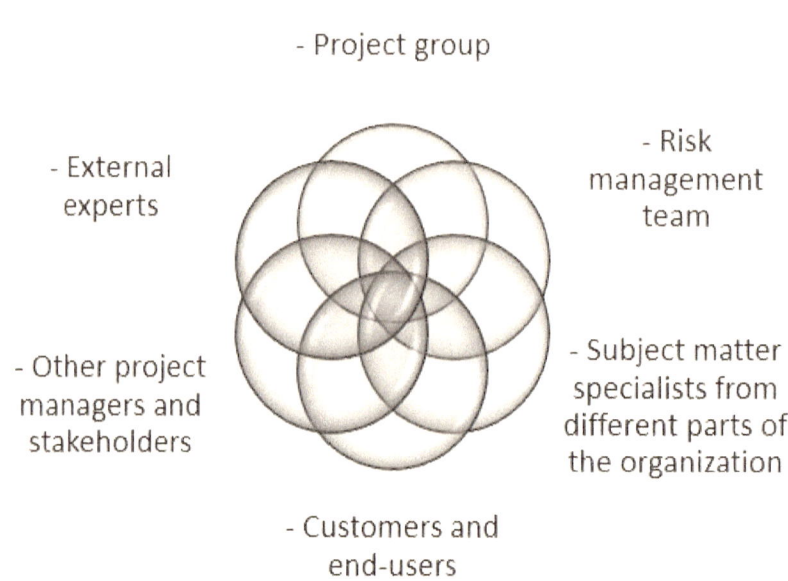

- Project group
- Risk management team
- External experts
- Other project managers and stakeholders
- Subject matter specialists from different parts of the organization
- Customers and end-users

The participants may differ however, the project group ought to dependably be included on the grounds that they are managing the project each day and, in this way, require new information as needed. External experts and stakeholders are essential as they provide information from an objective perspective.

All project stages must include risk identification process so as to avoid possible problems that may occur especially if such process is implemented once. The procedure begins in the initiation stage of the project where the initial risks are identified. In the planning stage, the project group decide risks with mitigation measures and ensures its documentation. In following phases of resource allocation, budgeting and scheduling the associated reserve planning is likewise recorded.

After the risks are identified in this initial phase, every one of the risks must be well managed till the project end. New risks will happen as the project develops and as the external and internal condition of the project changes, it serves as an ideal opportunity for the project manager to respond to it.

The project group and manager will be required to consider the arising challenges and suggest solutions to manage the problem. All these re-planning activities can mean a change to the budget, schedule and resource plan, which influences the completion of the project.

The risk management strategy should be clearly outlined at the early phases of the project with adequate documentation during the lifetime cycle of the project. The figure below reveals the risk management process.

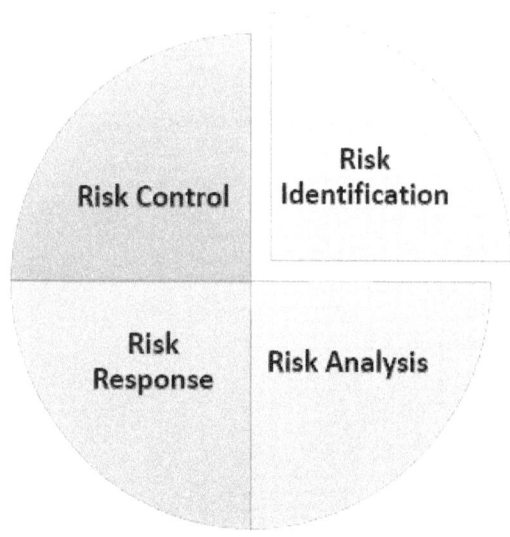

Figure: Risk Management Process

9.3 Risk Identification

Risk identification proof is the first and most imperative step since it serves as a fundamental step for subsequent ones. It consists of input at the initial step, techniques and tools in the middle and finally output at the exit process.

Figure: Risk Identification

For the primary input for risk, Examples of external factors are:

1. Economic conditions
2. Social, lawful or administrative patterns
3. Political atmosphere
4. Competition – Global or local
5. Fluctuation popular
6. Criminal or terroristic exercises
7. Internal factors sample are:
8. Internal culture
9. Staff abilities/numbers
10. Capacity
11. Systems and innovation
12. Procedures and procedures
13. Communication adequacy
14. Leadership viability
15. Risk appetite

Information from earlier projects, for the most part, records experience, improvements, clues, disappointments and risks of those previous projects which are currently valued in recognizing risks in the new project.

The lessons learnt from recent projects is an initial step for putting together organized information. Having an electronic copy of the same information is extremely helpful in the readiness of future projects. As you begin to audit this information toward the start of a project it might prompt positive ideas on how the project can improve or progress.

The Tools and Techniques step begins with the review documentation to effectively analyze existing information in written form. In project management it is the project plan and the planning documents:

1. Project charter
2. Project scope
3. Work breakdown structure (WBS)
4. Project plan
5. Cost estimates
6. Resource plan
7. Procurement plan
8. Assumptions list
9. Constraints list

The implementation of this techniques is trying to answer the question: "Is the project feasible or really realistic in terms of scope, budget and schedule?"

Methods used to identify risks related to projects

1. The risk breakdown structure: This shows a depiction of any known project risks, orchestrated by various classifications and their qualities in the vertical branches. Typically, it will demonstrate the greater part of the risks and their conceivable causes.

2. Brainstorming: This is a general information gathering session where creativity is utilized to identifying risks and conceive solutions to mitigating them. It is a meeting where subject matter specialists are "brainstorming" about risk sources and possible outcomes.

The thoughts are produced under the authority of a facilitator. The brainstorming meeting ought to be managed without criticism or judgments of ideas. More often than not, the ideas are built on other ideas. At the end, the risks identified will be sorted and properly sharpened.

The objective of brainstorming is to acquire a comprehensive list of project risks.

3. The SWOT analysis is likewise used to characterize conceivable risks. In the 1970s Albert Humphrey completed research at Stanford University utilizing information from 500 U.S. public corporations, and from this information, he built up the SWOT analysis.

SWOT is an abbreviation that searches for strengths, weaknesses, opportunities and threats. Regularly SWOT is utilized in brainstorming sessions. By defining strengths and opportunities, predictable weaknesses and threat, ideas do come to mind.

SWOT can be utilized for organizations, their various departments and divisions and also on an individual platform. A few benefits of using SWOT is that it is relatively cheap and simple to use. It also serves as a good platform to generate ideas.

However, it could consume time and may not often provide detailed information on how to reach a specific objective or reveal the importance of threat therefore, it is recommended that it be used carefully.

These techniques and tools aid the project manager to accumulate significant information to identify risks with opportunities and analyze it for the project, scope, cost and spending plan. This information will then be expressed on the supposed *risk report/register*, which is the major output of the risk identification step.

The risk register/report incorporates every recognized risk and their description, root causes of risks, risk categories, and the likelihood of occurrence, the single effects of specific risks, possible responses, and their causes.

The entire risk recognizable proof process has four principle sections on the risk enrol:

- Lists of recognized risks – Identified risks with their underlying causes and risk assumptions are recorded
- List of potential responses– Potential responses identified here will be the inputs to the risk response planning process
- Root sources of risk– Root causes of risk are the key reasons why we have the recognized risk
- Updated risk categories – The process of recognizing risks can prompt new risk classifications to be included

9.4 Risk Analysis

The foundation of risk analysis is the risk identification. Risk analysis covers an entire and consistent assessment which ought to be realized quantitatively as well as qualitatively for every single recognized risk.

The objective is to possibly identify interrelationships and enable the project manager to recognize a sort of significance order, likewise called prioritizing. It also helps to identify organizational goals and the consequences for the project. Risk evaluation should meet the following demands (Passenheim, 2009).

- Objectivity: A subject evaluation is carried out for internal risks while reference to the market is considered to ensure a practical objective for external risks.
- Comparability: A consistent and standard methods is to be employed by the organization to effectively evaluate risks which should lead to comparable results.
- Quantification: The organization is able to identify deviation from targeted objective by the means of quantification.

- Consideration of interdependencies: practically speaking, this is the hardest phase of risk evaluation. Not understanding the relationship between risks and their importance for the department and perhaps for the entire organization can be a major risk in itself.

That is the reason the project group ought to painstakingly consider the risk and the reaction it could mean for the group as well as for the entire organization since a good solution for one office can mean a problem for another office.

The most commonly applied method for analyzing risk is the **"scenario analysis"** which comprises of the probability of the event and the effect this would have on the project. The scenario analysis is part of numerous ways to deal with the analysis of risks, including mode and effect analysis (FMEA) or the program evaluation and review technique (PERT).

To properly conduct a risk evaluation the relevant levels should first be defined. For instance, the range should exist between 1 and 5 to give the effect or the probability a specific "size". And for more detailed evaluation, range 1 and 20 could be utilized.

Where more detailed information is required, a more exact classification of what the terms could signify as displayed in the figure below can be used. This could be depicted by letters and for likelihood or affected costs, percentages could be expressed for the distinctive evaluation levels.

Project Objective	1 Very Low	2 Low	3 Moderate	4 High	5 Very High
Cost	Insignificant cost increase	< 10% cost increase	< 10-20% cost increase	20-40% cost increase	>40% cost increase
Time	Insignificant time increase	< 5% time increase	< 5-10% time increase	10-20% time increase	>20% time increase
Scope	Scope decrease barely noticeable	Minor areas of scope affected	Major areas of scope affected	Scope reduction unacceptable to sponsor	Project end item is effectively useless
Quality	Quality degradation barely noticeable	Only very demanding applications are affected	Quality reduction requires sponsor approval	Quality reduction unacceptable	Project end item is effectively useless

Figure: Evaluation of Risks, adapted from (Passenheim, 2009) page 90

The evaluation form can be filled with the assistance of an expert. Estimation of the probable case is done as well as the worst and the best cases.

The FMEA (Failure mode and effects analysis) is a more detailed evaluation tool which will reveal the effect and probability by the detection possibility, which means that it is so difficult to really understand the occurring risk.

The enlarged equation with detection is:

Impact × Probability × Detection = Risk Value

The equation works by evaluating a five-point scale dimension. Detection shows the project team capacity identify that the risk is threatening. On the 1 to 5 scale, "1" would mean easily detectable "5" that the detection would most likely occur when it is considered late.

The result of the information would have a range between 1 and 125. "1" shows the risk has a low probability, a level 1 impact easy to detect. While that of "125" would demonstrate that the group need to deal with a high-impact risk and a high likelihood which could be identified.

Hence, helps us to take the decision whether to begin the project or not if the risk could not be transferred or mitigated. Considering all situation, the range between 1 and 125 can be utilized to define the dangerous nature of a risk.

9.5 Risk Response

It is possible that risk occurs after harvesting all data for the risk control. Therefore, the project manager will be required to demonstrate apt respond to it. The following five primary options can be utilized to best respond to such risk:

- Mitigate
- Avoid
- Transfer
- Share or
- Retain the risk.

Mitigation of risk is the lessening of the impact and the likelihood of risk occurrence. This is something one can do before beginning the entire project. Usually, an attempt is made to diminish the likelihood and after that the impact. The latter costs more to carry out but may not be relevant for consideration if the likelihood of occurrence could be highly reduced.

Prototyping and testing are common terms in engineering. By testing and prototyping, one can test the project on a smaller scale with less risk to detect possible problems and failure. The project team could use these techniques to better prepare for these problems and possibly eliminate them before commencing actual project work

Cost and time are two major things that cannot be mitigated easily. Since finance is spent and the days are numbered. However, risk management has an answer for this: **budget reserves and time buffers**. This is carried out with "safety ratio". This ratio is directly related to the experience gained from past projects.

Another common approach in risk transfer is **outsourcing**. Though the risk transfer process will cost money, the contractor takes the risk.

Additionally, a popular way to deal with transferring risks is **contracting insurance**. This may function properly for some particular cases yet for project management by and large it is not by any stretch of the imagination the correct approach. Contracting insurance for a project can be utilized for low-probability and high-impact events.

They are often costly for a daily business risk insurance but may be implemented for circumstances like earthquake or mudslide.

Sharing risk is another approach and as the name implies, different groups share the risks of a similar project. For instance, the Airbus corporation strategy – from the aircraft industry. Airbus designated risk to the R&D divisions over various nations like France, Britain and Germany.

Another sort of sharing of risk is marking a BOOT contract. BOOT is a shortened form for "Build-Own-Operate-Transfer", which means the project organization is building the plant, owns it to the point where operations run smoothly and the entire checkup is completed. Ownership is only transferred to the client when these steps are successful.

It is necessary to have a contingency plan else a risk may delay the response from the manager and any likely decisions made under pressure could be poor, expensive and possibly unsafe. The risk response matrix helps to follow up on risk related instructions and information.

9.6 Risk Control

Risk control is the last step of the entire risk management process. This step consists of executing the risk response methodology, monitoring and triggering events, initiating contingency plans, and constantly watch out for new risks.

The change management system is an essential part of risk control process and as the project progresses, the project manager must manage changes in scope, schedule and budget. The project group must be constantly conscious that unpredictable risks may happen.

However, this is not the standard case in each project. Team members are not continually eager to discover new risks and problems, especially when under great time pressure from top management to complete the project within a short time.

It is, therefore, the duty of the project manager to create an atmosphere where team members can freely raise concerns and admit their errors. External stakeholders and specialists ought to be brought into the discourse with the aim of reviewing the actual risk profiles of the project.

Another valuable key success factor is the task of assigning responsibility for every identified risk especially by the mutual agreement of all concerned stakeholders for proper communication to avoid certain risks being ignored.

Project auditing is also fundamental in the Risk Response Control. It helps to detect quality-related work and ensure that projects are being carried out as per standards and in line with the required plan. The major goal of auditing projects is to discover risks and weak points although this process could be time-consuming and a few employees could be offended seeing it as a form of control over their work and may dislike the auditor.

It is necessary to note that each audit process produces an audit report, for instance, the basic procedures and information for the observations and evaluation as far as project documentation or personnel capability. Obviously, non-conformities must be documented as well. This is essential for the following audit so as to check if corrective activities have been considered. Participants of the audit report must also be identified and recorded.

Change control is an essential part of the project. As the project develops someone must be in charge of approving changes, keeping the reports updated, and conveying all changes to the concerned stakeholders. Success depends vigorously on keeping the change control process updated.

Conclusion

It is clearly seen that if the project manager and all concerned personnel embrace risk management, it will enhance each project activity leading to the following improvements.

- Provides a system for implementing consistent and sustainable activities for the entire project
- Improves the basic leadership process, supports the planning procedure and prioritize every action by having a complete understanding of all project-related tasks, project opportunities and threats.
- Lessening uncertainty within the project
- Securing and or expanding the resources of the organization
- Optimizing operational effectiveness
- Developing and supporting the association's learning base

So, guys, always remember that risk management is a progressing process and you gain experience as you involve yourselves in multiple projects, stated Paul.

Mark questions, does this mean risk never rests?

Exactly, replied Paul has he sips his coffee.

QUESTIONS

CHAPTER 9: RISK MANAGEMENT

1. .... is the summation of opportunities and threats
   a) Certainties b) Threats c) Uncertainties
2. Risk management is a system to lessen the adverse effect of ....
   a) Financial gain b) financial loss c) all risk losses
3. Risk Identification is the project manager sole's responsibility
   a) True b) False c) No idea
4. Risk management process include .... and ....
5. Cost and time are two major things that cannot be mitigated easily
   a) True b) False c) No idea

ANSWERS

1. A   2. B   3. B.

4. Any of risk identification, risk analysis, risk controlling, risk response

5. A

CHAPTER 10

STAKEHOLDER MANAGEMENT

| Primary Project Stakeholders | Secondary Project Stakeholders | Managing Stakeholders | Understanding the interest and influences | Stakeholders and Communication |

10.0 STAKEHOLDER MANAGEMENT

Hi Paul, said Mark.

TGIF Paul, we are excited to be here today.

Fantastic, replied Paul. Managing Stakeholders is vital towards the success of any project guys.

In this chapter, we shall discuss stakeholder management by reviewing the following topics

1. Introduction
2. Primary Project stakeholders
3. Secondary project stakeholders
4. Understanding the interested and influences
5. Managing stakeholders
6. Stakeholders and communication

10.1 Introduction

Project managers must recognize the impact of their project on the external environment since every project will always relate to a larger environment which they exist. If this is neglected, project managers could face difficulties in the planning, approval and execution of the project.

The analysis of stakeholders gives the opportunity to easily identify necessary individuals who can influence or are influenced by the project. Edward Freeman in the early 1980s stated that a stakeholder in an organization is any individual or group of people who are affected by or can affect the achievement of the organization's goals and objectives.

This denotes that project stakeholders are persons or group of people affected by or can affect a project. Stakeholders can highly influence the project process.

10.2 Primary Project stakeholders

Primary project stakeholders are individuals whose influence on the project is almost instantaneous. That is, they can be rapidly influenced by the project or can influence the project in an instant. The individuals have a basic interest in the progress of the project due to their criticality in the very existence of the project (Smith, 2010).

These primary project stakeholders consist of sponsors, project champions, suppliers, equity and debt holders, contractors and staff on the project. They control the accomplishment or failure of the project since they are part of the project's processes.

Projects are not usually lone activities for primary stakeholders; they belong to a larger strategic activity. For example, the failure of a new product development project or technology can diminish the competitiveness or market standing of an organization. This reveals that the failure of such activity may affect the greater business activity.

It is fundamental that the project team fully comprehend the relevance of the project objective on the activities of project sponsors, champions or financiers. Although this may not come easy, as some stakeholders are not keen to reveal information that could be part of their competitive strategy. Nevertheless, it is vital to note that should a project be unsuccessful, not all primary stakeholders will be affected.

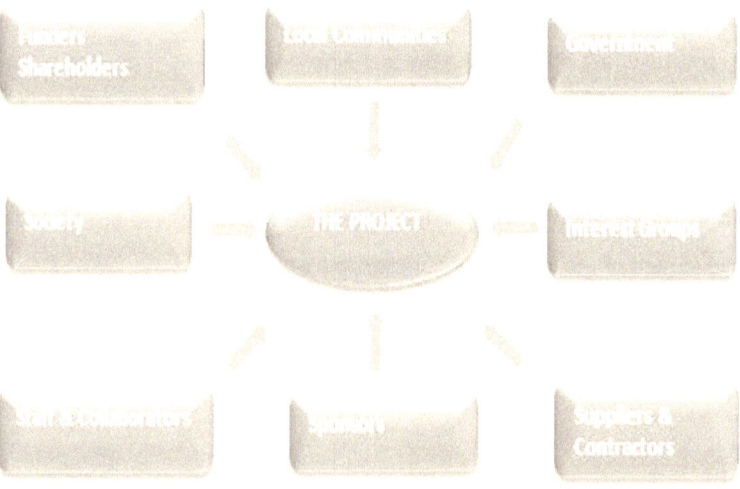

Figure: Primary and Secondary stakeholder map adapted from (Smith, 2010), page 129

10.3 Secondary project stakeholders

This category of stakeholders is not directly linked to the core of the project. They consist of parties such as the government, local authorities, unions, local communities, political parties, consumer groups and so on. Though secondary stakeholders are not directly related to the project, they can have a high level of influence on it.

For instance, policy, regulation and legislation of the government can exert a considerable level of influence on projects. Tougher environmental measures about disposal and waste management will also have a growing impact on engineering projects

The interest of the public to visible engineering projects or their by-products can either aid the promotion or disruption of projects. It is worth noting that secondary stakeholder influences can be felt during the feasibility and approval stages either at local or national level.

For approval processes in public arena, adequate consultation is required, and such activities take a while and could have a negative effect on the success of the project. Please permit me to share a quick example on the public enquiry of the proposed Terminal 5 facility at London Heathrow airport which took five years (Smith, 2010).

The expiry cost was expected to be approximately 100 million pounds. Various local, national and international interest groups including 11 local authorities were involved and were all against it. This resistance is expected to result in difficulty in ensuring that initial costs targets are archived.

The original business assumptions have changed including airline travel plan thus leading to a redefinition of some features of the business case for the project. Such a case reveals how engineering projects can suddenly nose dive or frustrated if there is opposition on any major scale by secondary stakeholders.

Although, the influence of secondary stakeholders can promote projects, in most cases it is disruptive. A basic factor in such feature is that secondary stakeholder usually has to live with the project outcome over its whole life-cycle contrast to the limited season primary stakeholders or project managers' experience.

10.4 Understanding the interests and influences

The word stakeholder is associated to the stake in the project. It deals the stakeholders interest, influence and impact with the project.

Specifically, the stakeholders influence the economic dimension of the project including the profitability, risk profile and cash flow of the project. In the same vein, the project can influence the stakeholders by its impact on market share, prices, growth and other factors resulting in gain or loss of fortunes depending on the project performance.

The technological area is another platform of influence the stakeholder could exert on the project. Some stakeholders may have the ability to inhibit or aid access to technology, equipment or skills by the project. This kind of situation could be prevalent if a highly-specialized technology is required and where stakeholder holds sway in a monopoly situation.

Another area of influence is the social arena. Stakeholders can exert this effect by changing public opinion hindering or promoting the company with their social influence. Project managers are usually unable to influence such circumstances unless they recognize the potential influence of stakeholders early in the project's life.

Societal influence depends highly on the morals, ethics and values present in any society, therefore, the value system of the sponsoring organization, its projects and its stakeholder will describe this relationship. Political influence powered by legislative capability could seriously impact projects as social issues often end up on the political agenda.

Organizations with high economic and political power can utilize their influence to alter political opinion to facilitate their projects and similarly disrupt projects. These stakeholders possessing political influence can redefine the potential impact of change on the company or organization.

It is thereby necessary that the political stake must be well-thought-out for sensitive nature projects or those which could have possible international impact.

A unique influence is also exerted in the managerial arena which is often generated internally from the project organization. This kind of influence deals with the management philosophy and culture of the organization thus, determining the systems, procedures, processes and values the company embraces.

Its impact is mostly felt on the existing relationship between the project and its internal stakeholders which create the internal culture of the project. A conducive environment of cooperation and trust will allow the project to thrive. It also allows for collaborative work relationships.

For instance, if a project is to be operated in a non-adversarial manner, the staff being recruited should demonstrate the ability to work in a manner that will fit into the culture of the project. It is to be noted that the value a company embraces internally will continue to reflect externally, hence a need to pay attention to the internal values

We have been able to categorize the multiple interests of the stakeholders to help us understand the nature of their relationship with the organization therefore, providing a mechanism to address any issues that may arise as the project advances.

10.5 Managing stakeholders

The changing nature and number of stakeholders and stakes as the project evolve is often a concern to the project manager. It is vital to identify the appropriate stakeholders as the project progresses continuously. This process of identifying potential stakeholders in a project should be carried out in the strategic management process of all projects.

Stakeholder management is an important part of strategic project management especially when we take the project as the equivalent of an organization. The process of aligning the objective of individuals who have an interest in the project is referred to as project stakeholder management.

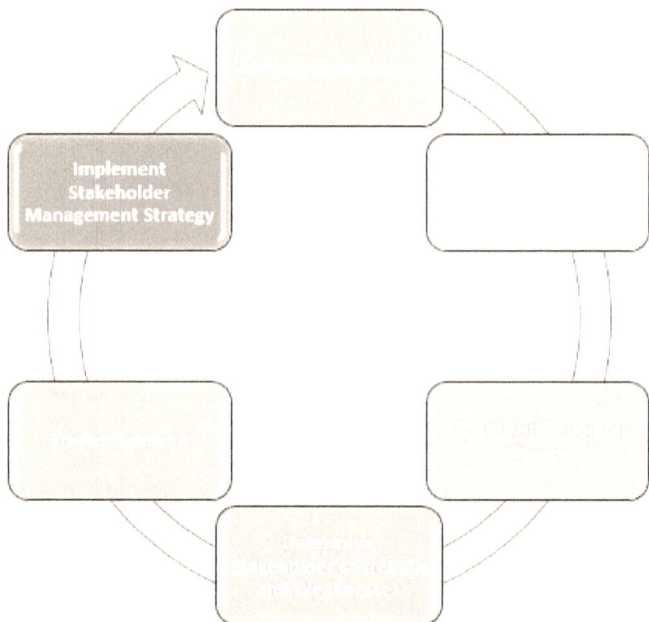

Figure: Project Stakeholder Management Process (adapted from Cleland and King, 1998)

In stakeholder management, the initial process is the identification of potential stakeholders on the project. The identification of primary and secondary stakeholders can be carried out with the simple stakeholder mapping.

To avoid future problems later in the project, creative and innovative thinking is required to identify stakeholders who have no obvious interest in the project.

The second step is to ensure all relevant information about the stakeholders is acquired as inadequate information could lead to an underestimation of the potential impact of stakeholders. Public organizations can easily access this information on the public domain but could be quite challenging for new organizations especially in a local context due to the difficulty that may face during research.

It is vital to recall that a minor well-organized group may constitute more challenges if they can successfully wield the power of the media or have adequate political influence.

Identifying the mission of the stakeholder for the project is the next step. For instance, if Yawdaw International is recognized as a potential stakeholder, acquiring relevant information such as how they work, what aspect of the project is likely to gain their interest and the impact the project may have on their company could be the foundation for determining the objective interest in the project.

It is important also to note that whereas the project manager and his team see opportunities from the project, other interests see threats.

The determination of the stakeholder weaknesses and strengths is the next stage. Primary stakeholders may be analyzed more easily as the desire the success of the project; however, the progress of projects could be frustrated if the secondary stakeholders are underestimated.

This step aids management to comprehend the influence stakeholders could have on a project, the available avenues to reveal their needs and how far they will go to see them through. The best scenarios are those of environmental interest groups that possess special organizations at all levels including international level.

They have a measure of influence on the public opinion, media, research institutes and even the political system. Nonetheless, they are often ignored in stakeholder analysis.

The attitude or behaviour of the stakeholder is to be anticipated and how this will affect the objectives of the project. The development of scenarios for various situations is highly recommended to establish possible contingencies and risk profiles for results.

Such scenarios help the project manager to implement stakeholder management strategies and respond effectively to possible negative attitude from stakeholder which result in actions that could affect or alter the project including additional resource allocation or effectively promoting project's objectives and ultimate benefits.

It is of importance to recall that interest in the project is necessary as it provides adequate support in circumstances where stakeholder attitude is positive nonetheless, both supportive and disruptive behaviour should be considered.

An example of how potential external stakeholders were identified by project managers to benefit a project is the Prince Edward Island Bridge project. The project managers on this project in Canada foresaw possible delays and consequent cost overruns due to the resistance of the fishermen and ferry companies.

To attract investors to the project, the promoters obtained a security package to guard against cost overruns and delays. Part of the package includes a US $ 141 million performance bond. This package aided the project to survive the seasons of delays as a result of court hearings and about 70 environment impact assessment studies that was carried out.

In this example, primary stakeholders assisted to provide solutions that neutralized opposition from secondary stakeholders. It is therefore fundamental to note that stakeholder management should improve the objective of the project achievement whereas the neglect of stakeholder inhibits it.

Stakeholder management process should also help to improve the motives and understanding of stakeholders.

10.6 Stakeholders and communication

The interest of the project should be effectively communicated to the project manager. This is necessary as the outcome of stakeholder identification reveals variation in interest of the audience in the project.

At this stage, communication is flowing to a wider audience from the project manager with information relating to progress, profitability, timing and performance which is conveyed to the primary stakeholders as their interest is tied to the project performance.

Communication strategies with primary stakeholders will include personal briefings, reports, newsletters, workshops, meetings, intranets and dedicated websites as this focuses on their specific interests.

Much of the information on the progress and performance of the project will be conveyed by the project manager to the primary stakeholders

In the case of secondary stakeholders, there are possibilities of communication challenges due to a variety of interests in the projects. This may vary from representing the project at a public inquiry to a newspaper interview about the project.

Stakeholder management allows the project manager to adequately recognize secondary stakeholders' interest and effectively channel communication to respond to these interests. It is quite vital that the project manager ensures that the communication includes the broader project objectives and benefits (Smith, 2010)

The project manager must not only concentrate on stakeholders interests as this may lead to losing sight of the ultimate project objective. Therefore, he/she must also employ relevant tools and skills to convey information which may differ from a presentation at a public meeting to a press release.

The project manager must also manage negative communication or misinformation using rebuttal techniques. It is possible that secondary stakeholders resist project progress by waving negative information and media campaign to favour their interest especially with the use of the internet.

The project team must prepare for possible disruption by direct action, protests, strikes and boycotts. Although stakeholder analysis should identify the likelihood of such circumstances and adequate contingency plans be made to tackle this.

Furthermore, in situations where projects face more scrutiny, project managers should procure advanced communication skills, and when project team members are confronted with advanced and complicated campaigns, the aid of a specialist could be implored from either sponsors or external media specialists.

Thanks so much for this knowledge Paul, commented Mark.

We must never overlook the influence stakeholders may have on our projects, remarked Bill.

Correct, responded Paul.

QUESTIONS
CHAPTER 10: STAKEHOLDER MANAGEMENT

1. Stakeholders can highly influence the project process.
   a) True b) False c) No idea

2. The process of aligning the objective of individuals who have an interest in the project is …..

3. identification of potential stakeholders on the project is the third process in stakeholder management
   a) True b) False c) No idea

4. The project manager must only concentrate on stakeholders interests to have a successful project.
   a) True b) False c) No idea

5. Stakeholder management allows the project manager to recognize stakeholders interests and channel communication to respond to these interests
   a) True b) False c) No idea

ANSWERS
1. A   2. Project stakeholder management   3. B   4. B   5. A

CHAPTER 11

PROJECT CLOSURE

> Testing and Start-Up

Final Inspection

Guarantee and Inspections

Planned Drawings and Records

> Disposition of Project Plans

> Post Project Critique

Client feedback

11.0 PROJECT CLOSURE

Congratulations guys, we are approaching the end of this course, commented Paul.

Thank you sir, it has been really impactful, responded Mark and Bill

So, at this stage, we shall discuss the strategies to bringing the project to a successful close. That is, how will the engineer project manager wrap up the project effectively? We shall be covering the following areas in this chapter

1. System Testing and Start-Up
2. Final Inspection
3. Guarantee and Inspections
4. Disposition of Project Plans
5. Post Project Critique
6. Owners Feedback

11.1 Testing and Start-Up

For overwhelming modern plant projects, investigation of construction is performed all through the project nonetheless, the client or his representative by and large requires review of all products or items before conclusion and the testing of key machines and equipment upon establishment.

The term 'mechanical completion is regularly used to characterize the phase of project development in which these strategies are performed. Sometimes hard to characterize mechanical completion; thus, the project manager ought to build up a plant finishing standard with the client to clarify what constitutes conclusion (Oberlender, 2000).

This ought to be in the construction contract with the goal that all required personnel in the project know who is to do what for each period of the project. The project manager will have to work with the construction contractor to facilitate the interface between the client, the important originators, and the contractor. The responsibilities and tasks of everyone must be unmistakably characterized.

Additionally, the classes of tests and the methodology for testing must be plainly described as per the contract documents.

A formal arrangement is to be prepared when a machine can be shut, the lead time notice required for investigation, what is to be observed, and a close sheet for the client. This is required to wipe out pointless opening and shutting of machines, which could require an ample proportion of project time.

The project manager must acquire in writing from the client's representative the procedure for turning machines over to the client. Effective custody and care are significant because considerable expenses are included. It is imperative for each project team member to know who is responsible and when they are responsible.

The project manager ought to inform the plant when a specific equipment or machine is finished, tried, and prepared for turnover to the owner or client. Upon acknowledgement, any extra changes require a work approval from the owner.

This procedure is to be implemented formally and include signatures from the leader and responsible representatives. The project manager needs to arrange with the contractor and designer to define the procedure or strategy for a start-up.

The procedure must be formal, nonetheless flexible. The project manager should also acquire in writing from the client's or owner's representative what they require from the different individuals in the project group for support amid start-up. The project manager must consult owner for his input and suggestion and not assume unnecessarily.

So many responsibilities for the project manager, commented Bill

Yes, and that is why he must be on top of his game, replied Paul.

11.2 Final Inspection

Assessment of construction work is performed all over the span of the project. As the project is implemented, equipment, machines, electrical frameworks, and mechanical systems might be completed and prepared for testing and acknowledgement as per the contract agreement.

The project manager must work intimately with the owner's representative and the design experts who oversee the investigation, testing, and final acceptance.

A definition of mechanical completion is to be produced and a formal notification given to the construction contractor that permits sufficient lead time for the procedure. This is vital so significant time is not lost, which may unfavourably influence the project fruition date.

There must be an unmistakable understanding to what the owner needs to check during tests, what tests they need to witness, and the required testing type. The responsibilities of the three foremost contracting parties (the client, contractor, and designer) must be adequately identified. It is the obligation of the project manager to organize this effort successfully.

The commencement of project close out begins as the project comes to fruition or completion when the contractor asks for a last examination of the work. Preceding the demand, a punch rundown list is prepared to identify all things may require remedy.

To build up this punch list, the field investigation faculty personnel must precisely review their daily inspector's log to note all work items which have been entered that require remedial activities. It is often important to recycle through the punch list process a few times before the work is palatable for acknowledgement.

The final stroll of inspection should incorporate representatives of the owner, contractor, and the key design experts (the architect, along with civil, electrical, mechanical, project engineers and so on.) who have dealt with the project. The project manager should schedule and conduct the final stroll inspection.

Acknowledgment and acceptance of the work including the final payment to the contractor must be done as per the details in the contract agreement. The project is said to be substantially completed when construction has been adequately carried out as per the contract documents, and the project can be utilized for the reasons it was intended.

This implies that only minor items are left to be completed and that the project is sufficiently finished to be utilized. The contractor may issue a Certificate of Substantial Completion with a joined rundown of all work left to be done to finish the project.

Approval of the Certificate of Substantial Completion, with the appended deficiency list, affirms acknowledgement of the completed project work. In this way, it is vital to guarantee that the rundown list is completed as the contractor has no further commitments under the contract after the owner signs the certificate.

Usually, final instalment or payment, including the release of all retainage, is withheld for thirty to forty days after fulfilment of all deficiencies. Before final payment, the contractor is to submit last paperwork: guarantees, warranties, lien releases, and other related documents as required by the contract.

11.3 Guarantee and Inspections

For the most part, the contract requires the contractor to ensure all materials, machines, equipment, and work to be of good quality and free of deformities in line with the contractual agreement for a time of one year after construction is completed.

The warranty of the general project can be reached out past the ordinary one-year time frame; notwithstanding, this is not the case most times.

Distinct machines or equipment frequently have guarantees that stretch out from one to five years after commissioning. Working guidelines, manuals, spare parts, and warranty declarations must be provided to the owner. The project manager is to ensure that all guarantees are compiled and provided to the owner before the last installment to the contractor.

11.4 Planned Drawings and Records

It is usually expected that modifications and changes occur during project implementation. A minimum of one set of the original contract agreement issued for bidding purposes must be kept for reference.

This is essential for the resolution of claims and conflicts which often arise for example, "What did the contractor bid on? Moreover, there must be a careful documentation of all change-orders amid construction.

A typical contract necessity is that the contractor must have a copy of all specifications, drawings, addenda, change-requests, and shop drawings that may not have been in the initial plan. These 'as-built' drawings also reveal measurements and work details that were not performed as they were initially shown.

Examples are changes in entry and exit points to buildings, the layout of electrical wires, cooling channels, or the area of underground piping, utilities, and other concealed work. These documents reveal every change to the original contract agreement and are given to the client or owner as the project is completed.

11.5 Disposition of Project Plans

As the project is being executed, the project manager regularly maintain two documents:

A recorded document and a working document. The recorded document contains original copies of vital data about agreements, contracts, and other legitimate matters.

The working document is utilized by the project manager for daily administration of the project and normally contain record copies of correspondence, minutes of meetings, phone logs, reports, and the likes. When the project is completed, a substantial amount of information related to the project, including the records and documents are collected.

Most firms or organizations have a characterized method for document disposition. Data from the record document ought to be well organized and structured for easy accessibility for future reference.

The working document usually contains duplicated information record, some of which may have been manually written including notes, such data should not be wiped out. However, a great part of the information could be disposed. Adequate information should hold for easy retrieval by the project manager to enable easy follow up on his or her work on the project.

11.6 Post Project Critique

A post project critique or review is to be carried out as soon as the project comes to an end as this enables learning, i.e. information and experience gained from each project that can be utilized to enhance the achievement of future projects. Participants to be incorporated should include the client/ owner, lead designers, team members, and construction representatives.

The feedback received through a non-accusatory discourse of the challenges and solutions experienced amid a project implementation is valuable and beneficial to all colleagues in the strategic planning and execution of future projects.

For the meeting to accomplish the intended outcomes, it is essential that the discussion is carried out with motivation and professionalism. All the project areas should be covered in the discussion including both the good and bad areas.

The purpose of this is to emphasize how to maximize the feedback information to reduce possible problems encountered in the future unlike blaming team members for the bad areas of the project. Meeting minutes are to be adequately conveyed to individuals absent at this post project evaluation phase. This will enhance knowledge sharing and profit everyone by the lessons learned.

A project peer review could also be implemented addressing the needs of the client, designer or another concerned party.

11.7 Client feedback

The importance of feedback from the owner or client cannot be over emphasized since the key determinant factor of a successful project is how adequately it met the objective of the client and how well it was utilized by the client. Therefore, after a project is completed and in use by the owner, a formal meeting should be held with representatives from the owner's organization to obtain feedback regarding the performance of the project. This is an important activity for evaluation of the quality of a completed project and satisfaction of the owner.

What a brief session today, commented Bill.

Yeah brief and enlightening, responded Mark.

QUESTIONS
CHAPTER 11: PROJECT CLOSURE

1. It is not expedient to review and test machines and equipment upon establishment of a project
   a) True b) False c) No idea

2. The project manager must orally acquire the client's representative procedure for turning machines over to the client
   a) True b) False c) No idea

3. The project closeout begins as the project comes to fruition or completion
   a) True b) False c) No idea

4. A post-project review is to be carried out as soon as the project is completed to enable learning
   a) True b) False c) No idea

5. Client feedback is an important activity for evaluating the quality of a completed project
   a) True b) False c) No idea

ANSWERS
1. A   2. B   3. A   4. A   5. A

CHAPTER 12

FUTURISTIC VIEW

ENGINEERING PROJECT MANAGEMENT – THE WAY FORWARD

12.0 FUTURISTIC VIEW

This final chapter unveils the way ahead for engineering project management. Paul stated.

12.1 ENGINEERING PROJECT MANAGEMENT – THE WAY FORWARD

It could be subjective to correctly predict the development and evolution of existing systems, especially in management. This is due to the likelihood of seeing less as we target the future, however, to make progress, we must attempt the right prediction.

To predict accurately, the analysis of current situation and the related trends in history is to be successfully utilized as a mechanism to project into the future. This method may be acceptable for short term predictions and technological processes. Nonetheless, it is limited in its capacity to satisfactorily cover the wider view.

As we evolve in time, newer techniques and processes spring up and are promoted in different degrees of success including business process reengineering and total quality management. Although such processes can stand alone, however, they are viewed as part of the overall project management processes.

In the same vein, there is likelihood to further develop innovative and novel processes which will be incorporated into best practice for the delivery of world class project management. The above techniques increase project effectiveness since project management relies upon good management practice.

Interestingly, these techniques achieve this by focusing on non-project parameters. The most prosperous of these individual techniques are currently utilized by an organization not involved in project management for their internal and external projects. Hence, it is possible that in the coming years, Project management will incorporate these non-project management approaches.

Although, the integration of discrete business function to improve the effectiveness of decision making to enhance project management techniques, it is likely to include some changes in terminology and approach.

Nevertheless, until such time, project management will continue to ensure that projects are delivered to target the requirements of the promoter and other possible project constraints, initially and always. Using the current performance to measure and effectively predict, there is a likelihood that project management will continue to be relevant for the future (Smith, 2010).

It is the opinion of the author that Project Management Engineering (PME) will remain relevant in years to come. Though, it may find expression in different forms, the fundamental principles of managing projects may not change much but will undoubtedly evolve with the arrival of Artificial Intelligence (AI) and internet at the speed of light.

Thanks for your attention and time, commented Paul.

The honor is all ours, replied both Mark and Bill.

REFERENCES

Constructability Concepts File, Publication No. 3-3, Construction Industry Institute, Austin, TX, August 1987

D.I Cleland and W.R King. (1998). *Project Management Handbook.* New York.

Elbeltagi, D. E. (2009). *Construction Management.*

Heerkens, G. (2002). *Project Management.* McGraw-Hill.

John M. Nicholas and Herman Steyn. (2012). *Project Management for Engineering, Business and Technology.* Abingdon, Oxon: Routledge.

Kam Shadan and Gannett Fleming. (March 2012). Construction Project Management Handbook. Washington DC 20590, 1200 New Jersey Avenue, SE , U.S. Department of Transportation. Retrieved from http://www.fta.dot.gov/research

Munroe, M. (1997). *Understanding the Purpose and Power of Woman.* Destiny Image.

Nevitt, P. (1983). *Project Finance.* Bank of America, Financial Services Division .

Oberlender, G. D. (2000). *Project Management for Engineering and Construction.* Thomas Casson.

Passenheim, O. (2009). *Project Management.* Olaf Passenheim & Ventus Publishing ApS.

Rita Mulcachy, PMP, et. al. (2013). *PMP Exam Prep.* RMC Publications, Inc.

Smith, N. J. (2010). *Engineering Project Management.* Oxford, UK: Blackwell Publishing Company.

Technologies, I. (n.d.). *Practical Project Management for Engineers and Technicians.* Retrieved from www.idc-online.com

About the Author | Oluwaseun Adenigba

Oluwaseun is a Fellow member of American Academy of Project Management and a member of Nigeria Society of Engineers.

He is an alumni of globally recognized institutions including Covenant University, Ota, Nigeria, Cardiff University, Wales, United Kingdom where he bagged his B.Eng. and MSc. in Mechanical and Advanced Mechanical Engineering respectively. He also has a Diploma in Financial Management from the prestigious International Business Management Institute, Berlin, Germany.

Some of his certifications include Certified International Project Manager (CIPM™), CompTIA Project+, Project Management Essentials Certified, and Master Certificate in Business Management.

A certified Life Coach skilled in Capability Development, Engineering Management, Inventory Management, Procurement, Negotiation & Conflict Management, Advanced Customer Service, Coaching and Public Speaking.

His global impact emanates from Nigeria where he is based and can be reached at adenigbaoa@icloud.com or Yawdawinfo@gmail.com.

Oluwaseun is a family man and author of *Mastering the Art of Project Management Engineering* with two other books.

NOTES

NOTES

Detail Qualifications

ACADEMIC QUALIFICATIONS

Dipl. Financial Management

International Business Mgmt. Institute, Germany

MSc Advanced Mechanical Engineering

Cardiff University, Wales, United Kingdom

BEng (Hons) Mechanical Engineering

Covenant University, Ota, Nigeria

PROFESSIONAL CERTIFICATIONS and MEMBERSHIPS

Professional Certifications

Life Development

- Life Coach – International Association of Professions Career College

Technical Development

- Lean Six Sigma Primer – Campus Business Management Group
- Managing Supplier Performance – Next Level Purchasing Association

Project Management

- Certified International Project Manager – American Academy of Project Management: 2018
- Project Management Essentials Certified – Management & Strategy Institute: 2017
- Certified Project Manager - CompTIA Project+ :2012

Health and safety

- Occupational Safety and Health - OSH Academy Safety and Health: 2014

Management and Leadership

Via Master Class Management:

    Master Certificate in Business Management

Via United States Institute of Peace:

    Negotiation and Conflict Management

    Conflict Analysis Course

    HR Managers Capacity Development

Sustainability and Environment

Via United Nations Information Portal on Multilateral Environmental Agreements:

- Introduction to Environmental Governance
- Introductory Course to the International Plant Protection Convention
- Introductory Course to the United Nations Convention to Combat Desertification

Publications

- Mastering the Art of Project Management Engineering
- Easy Guide to understand how MATLAB works: *Transitioning from Basics to Advance (Co-author)*

Sites Published
https://www.amazon.com/s?k=Oluwaseun+adenigba&ref=is_s

Location: Lagos, NG:

Optimization of a **10 million USD** worth of spare parts inventory.

PROJECT 1: COST REDUCTION of MACHINE SPARES

Objective: Reduce Cost of Ordering Spare Parts by 50%

Role Occupied: Assistant Project Manager

Achievement/Lesson Learnt

Reduction of extra charges on spares ordering was reduced by 95% resulting in a monthly reduction expense from **$9000 to $300.**

PROJECT 2: EXPANSION of ENGINEERING STORES ROOM

Objective: Decongest Current Store room by 30%

Role Occupied: Assistant Project Manager

Actual Work done

The new store was to accommodate inventory spares which worth about **2.5 million USD equivalent to about 1 billion naira**.

Achievement/Lesson Learnt

Decongestion of store by **40%** resulting faster issuance and receipt of spares

Creation of jobs for six store personnel to manage new store

Easier tracking of spares in store due to increased visibility

Positive feedback from maintenance engineers on store operations

www.ingramcontent.com/pod-product-compliance
Lightning Source LLC
Chambersburg PA
CBHW021411210526
45463CB00001B/320